Biotechnology Monographs

Volume 4

Editors

S. Aiba · L.T. Fan · A. Fiechter · J. Klein · K. Schügerl

K. Nakamura   M. Aizawa   O. Miyawaki

Electro-enzymology Coenzyme Regeneration

With 79 Figures

Springer-Verlag
Berlin Heidelberg New York
London Paris Tokyo

Prof. Dr. Kozo Nakamura
Dept. of Chemical Engineering
University of Gunma
Kiryu, Gunma 376
Japan

Prof. Dr. Masuo Aizawa
Dept. of Bioengineering
Tokyo Institute of Technology
Meguro-ku, Tokyo 152
Japan

Dr. Osato Miyawaki
Dept. of Agricultural Chemistry
The University of Tokyo
Bunkyo-ku, Tokyo 113
Japan

ISBN-13:978-3-642-73126-6   e-ISBN-13:978-3-642-73124-2
DOI: 10.1007/978-3-642-73124-2

Library of Congress Cataloging-in-Publication Data. Nakamura, K. (Kōzō), Electro-enzymology, coenzyme regeneration/K. Nakamura, M. Aizawa, O. Miyawaki. (Biotechnology monographs; v. 4) Includes bibliographies. 1. Enzymes–Electric properties. 2. Enzymes–Industrial applications. 3. Coenzymes–Industrial applications. 4. Bioreactors. I. Aizawa, M. (Masuo). II. Miyawaki, O. (Osato). III. Title. IV. Series. QP601.N26   1988   660.63–dc 19
ISBN-13:978-3-642-73126-6 (U.S.)

© Springer-Verlag Berlin Heidelberg 1988
Softcover reprint of the hardcover 1st edition 1988

2152/3145-543210 Printed on acid-free paper

Preface

As a key technique in enzyme engineering, development of the immobilization of biocatalysts has contributed to the new industrial utilization of enzymes and microbial cells during the past two decades. Many products such as amino acids, penicillin, and high fructose syrup have been produced in bioreactors with the immobilized enzymes or microbial cells. Other examples of industrial applications of biocatalysts will be realized in the production of chemicals, pharmaceuticals and foods. The biocatalyst with a high substrate specificity is also useful in the field of analytical chemistry, especially in the analysis of a special compound in a mixture with no need for a purification step. Biosensors are the representative example in this category. In this monograph, two topics were chosen, i.e. electro-enzymology and coenzyme regeneration. Both of them, we believe, are related to key technology in the future development of enzyme engineering.

There are several enzymes and coenzymes which are electrochemically active. Therefore, electrochemical techniques are useful for the electrochemical characterization of these compounds. More importantly, some enzyme reactions are capable of being coupled with electrochemical processes. This is applicable to electrochemical regeneration of coenzymes and electrochemical control of enzyme activity. Electroanalytical chemistry and bioelectrical energy conversion are other fields of the application of the coupling of enzyme reactions with electrochemical processes. Chapter 1, written by MASUO AIZAWA, is devoted to these topics under the title of electro-enzymology.

The industrial application of the biocatalysts has been increased, while the types of the enzymes used are limited only to hydrolases, isomerases and lyases. Usage of other types of enzymes such as oxidoreductases, transferases and ligases should be explored because these enzymes have strong potentials for the synthesis of chemicals and pharmaceuticals. As many of these enzymes require expensive coenzymes, coenzyme regeneration will be the prerequisite for the application of these potentially useful enzymes. Details of the coenzyme regeneration, that is, methods, processes, chemically modified coenzymes, and bioreactor systems are described in Chapter 2, written by OSATO MIYAWAKI and KOZO NAKAMURA.

Each chapter, based on about three hundred references involving author's original work, is intended to give readers a clear and authoritative overview of the present status of each field. Tables and figures included in abundance will be helpful for this purpose.

We would like to express our sincerest thanks to the editorial committee of "Biotechnology Monographs", particularly to Emeritus Prof. SHUICHI AIBA,

University of Osaka, who provided us with the opportunity for writing this monograph. We also appreciate the friendly cooperation of the staff of Springer-Verlag.

Tokyo, July 1988

KOZO NAKAMURA
MASUO AIZAWA
OSATO MIYAWAKI

Table of Contents

1 Electro-Enzymology

1.1 Introduction

Electro-enzymology is an interdisciplinary field resulting from the fusion of electrochemistry and enzymology. Electrochemistry and enzymology have traditionally been regarded as separate and distinct bodies of knowledge, each with its own set of laws and principles. Electro-enzymology has, however, raised a number of proximities between them.

Important stimuli for the dramatic development of electrochemical methodology were made from the mid 1950s into the early 1970s. Sophisticated electronic instruments have allowed for greatly increased experimental flexibility. It is also noted that widespread use of computer technology contributed to data analysis and theoretical treatments. Now electrochemists have turned most of their atten-

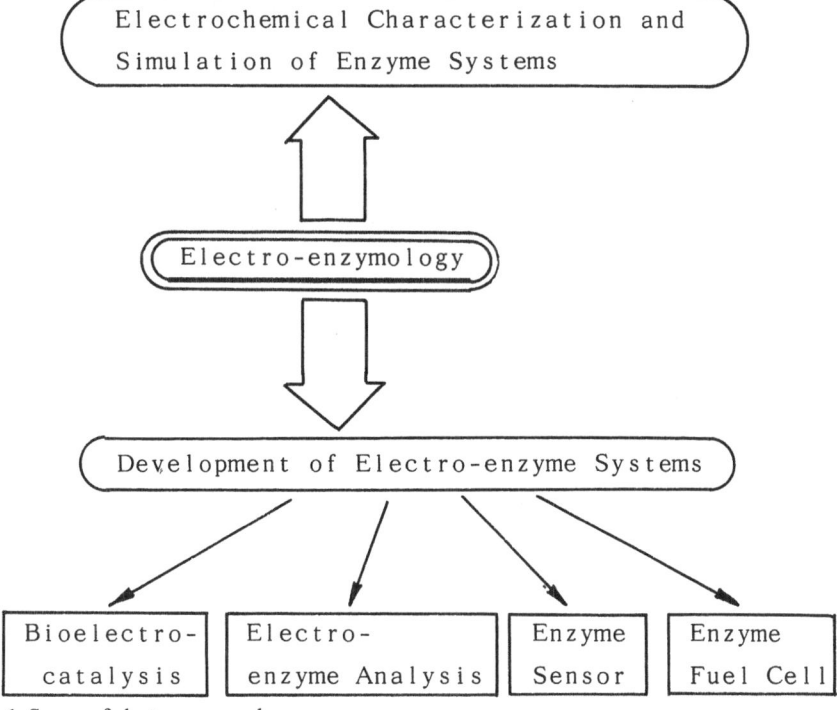

Fig. 1.1. Scope of electro-enzymology

1

tion to chemical, especially biochemical problems, rather than the methods themselves, and the methodology has become a separate body of knowledge.

The endeavor to make practical use of enzymes has been supported by continuing developments in the field of enzymology within recent years, during which the structures and mechanisms of action of various enzymes have been elucidated. The question necessarily arises of how to exploit this knowledge in practice. Of the many possible practical applications of enzymes, one promising approach has been the consideration of electrochemical methods in combination with enzymes. These methods now have been successfully introduced to enzymology, substantially contributing to enzyme characterization and the elucidation of reaction mechanisms. Some electrochemical processes have been linked with enzyme reactions leading to what is considered as: bioelectrocatalysis. Moreover, enzyme sensors and enzyme fuel cells utilize the exceptional properties of enzymes to recognize specific substrates and catalyze specific reactions. The further development of electro-enzymology will certainly lead to new applications of enzymes in various fields.

This chapter presents electro-enzymology in the scope of Fig. 1.1 and in the form of bibliographic review.

1.2 Electrochemical Characterization of Enzymes and Coenzymes

Although living organisms may appear to be in equilibrium, because they may not change visibly as we observe them over a period of time, actually they exist in a steady state, in which the rate of transfer of matter and energy from the environment into the system is exactly balanced by the rate of transfer of matter and energy out of the system. A living cell therefore is a nonequilibrium open system, a molecular machine for extracting free energy from the environment, which it causes to increase in entropy. Another reflection of the principle of maximum economy is that, living cells are highly efficient in handling energy and matter. The chemical reactions and processes of cells have been refined far beyond the present-day capabilities of chemical engineering.

It is therefore worthwhile to elucidate the mechanisms of these molecular assemblies in the living organism. Electro-enzymological investigations have focused on biological electron transport especially in respiration, photosynthesis, and nitrogen-fixation; various electrochemical methods have been employed to characterize each component involved in these reactions.

This section describes the electrochemical properties of enzymes and coenzymes involved in biological electron transport.

1.2.1 Redox Potentials

1.2.1.1 Definition and Methodology [1–3]

Measurement of redox potentials is important for understanding the intermolecular electron transfer between proteins or between substrates and enzymes, and the

2

intramolecular electron transfer between different redox centers within one macromolecule.

The standard potential, E^0, is a thermodynamic value that can be measured potentiometrically for ideally reversible redox couples that rapidly reach equilibrium. The Nernst equation, Eq. (1.1), defines the standard potential, E^0, on the basis of the activities of the oxidized and reduced forms, a_{ox} and a_{red}:

$$E = E^0 + (RT/nF) \ln (a_{ox}/a_{red}), \tag{1.1}$$

where n is the number of electrochemical equivalents, R the gas constant, and F the Faraday constant. The generally used formal potential, $E^{0\prime}$, is based on the

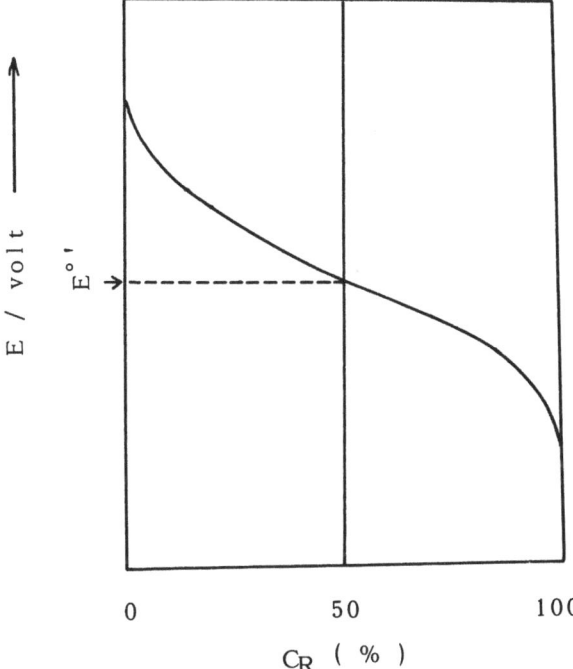

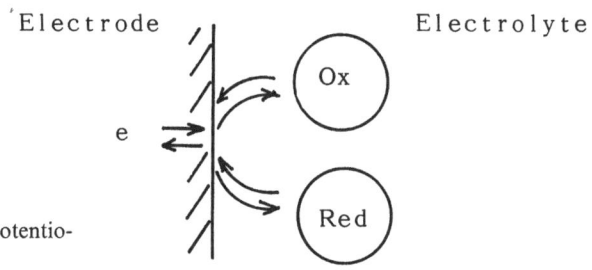

Fig. 1.2. Principle of potentiometric titration

3

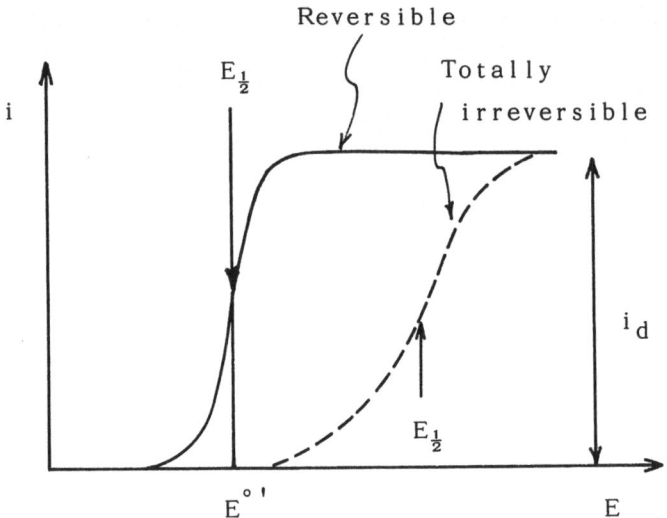

Fig. 1.3. Polarographic determination of redox potential

concentrations of the partners in the redox reaction:

$$E = E^{0\prime} + (RT/nF) \ln(C_{ox}/C_{red}),\qquad(1.2)$$

where C_{ox} and C_{red} are the concentrations of the oxidized and reduced forms. Formal potentials may be determined by the following methods.

  (i) Potentiometric titration: At half conversion, i.e., when $C_{ox} = C_{red}$, the measured potential is $E^{0\prime}$ according to Eq. (1.2), as schematically shown in Fig. 1.2.
 (ii) Polarography: The following relation, Eq. (1.3), holds for the half wave potential, $E_{1/2}$, determined by DC polarography for a reversible couple:

$$E_{1/2} = E^{0\prime} - (RT/2nF) \ln(D_{ox}/D_{red}),\qquad(1.3)$$

where the D terms refer to the diffusion coefficients of the oxidized and reduced species. In most cases, the polarographic half wave potential, $E_{1/2}$, of a reversible system corresponds to the formal potential as shown in Fig. 1.3.
For an irreversible electrode reaction, Eq. (1.4) holds:

$$E_{1/2} = (RT/nF) \ln(k^{0}t^{1/2}/0.76\, D_{ox}^{1/2}),\qquad(1.4)$$

where k^{0} is a symmetry factor and t is the polarographic drop time.
(iii) Cyclic voltammetry: The peak potential, E_{P}, correlates with $E_{1/2}$ as described by Eq. (1.5) and as expressed through Fig. 1.4.

$$E_{P} = E_{1/2} - 1.11\, RT/nF.\qquad(1.5)$$

4

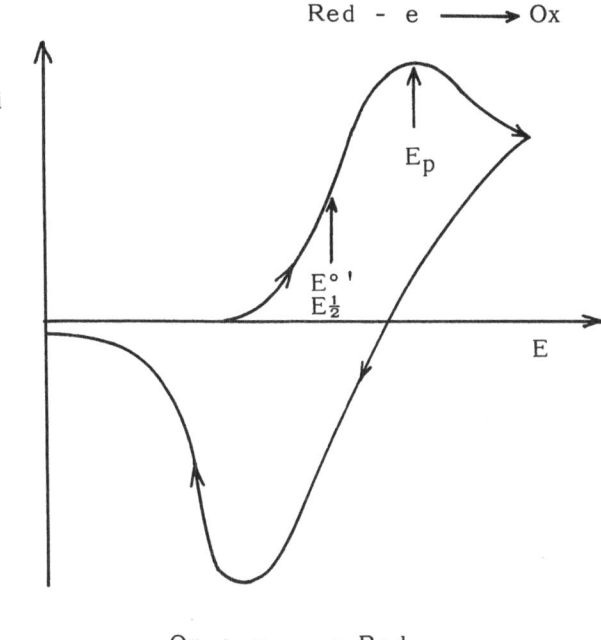

Fig. 1.4. Correlation of peak potential and half wave potential

(iv) Optically transparent thin-layer electrode: These are thin films of a semiconductor (SnO_2 or In_2O_3) or a metal (Au or Pt) deposited on a glass, quartz, or plastic substrate; or fine wire mesh "minigrids" with several hundred wires per centimeter. The optically transparent electrode is installed in a thin-layer optical cell, with a semi-infinite linear diffusion of electroactive species to the electrode surface.

The absorbance change can be described by considering a segment of solution of the thickness dx and a cross-sectional area a. If a species "red" is the only species absorbing at the monitored wavelength, the total absorbance becomes:

$$Ab = \varepsilon_{red} \int C_{red}(x, t) dx, \tag{1.6}$$

$$Ab = 2 \varepsilon_{red} C_{ox}^* D_{ox}^{1/2} t^{1/2} / \pi^{1/2}, \tag{1.7}$$

where ε_{red} is the molar absorbance [1].

Equilibrium is achieved rapidly, since the diffusion path is short. The equilibrium ratio C_{ox}/C_{red} at an applied potential is thus measured spectroscopically as illustrated in Fig. 1.5.

(v) Coulometric titration with mediators: The titrant is electrochemically generated by the reduction of a low-molecular-weight mediator. The reduced mediator then transfers electrons to the protein in the solution until equilibrium is reached [Eq. (1.9)]. The redox potential of the mediator (E_M^0) has to be close (within 180 mV) to that of the potential E_P^0. The mediator is most effective

5

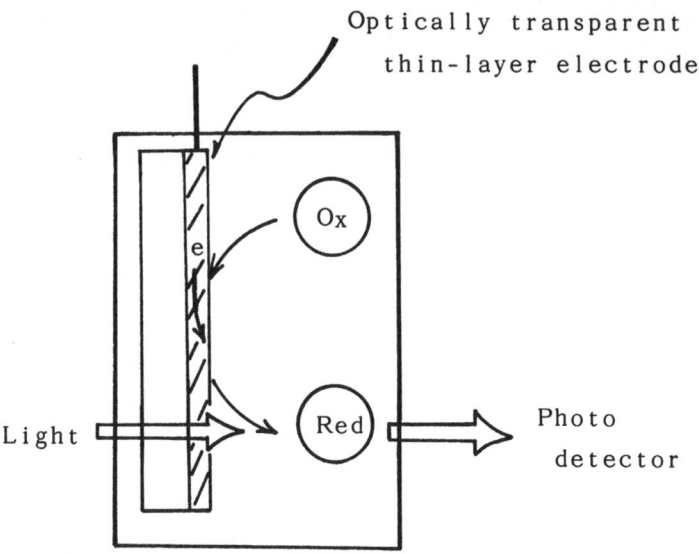

Fig. 1.5. Determination of redox potential with an optically transparent thin-layer electrode

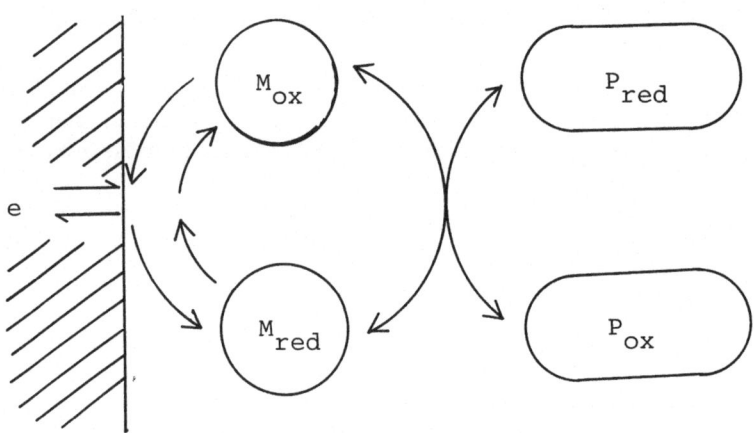

$$E^0_P - E^0_M = RTlnK$$

Fig. 1.6. Coulometric titration with mediators

Table 1.1. Redox potentials of electron mediators

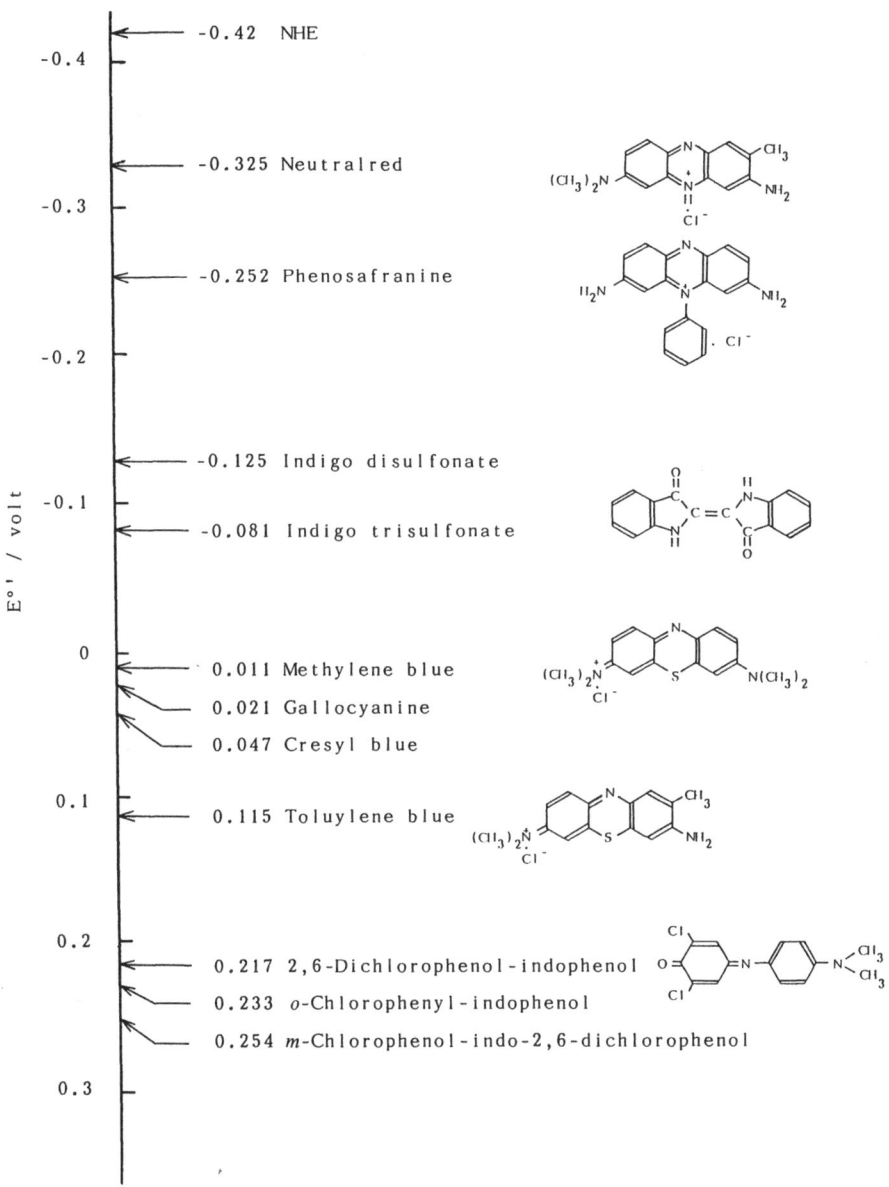

near a 1:1 ratio of oxidized to reduced form. Electrons readily transfer from the protein to an electrode via the mediator to establish equilibrium. Several mediators are listed in Table 1.1. The process for the coulometric titration with mediators is illustrated in Fig. 1.6

Electrode reaction

$$M_{ox} + ne \rightarrow M_{red},$$ (1.8)

7

Solution reaction

$$M_{red} + P_{ox} \rightarrow M_{ox} + P_{red}.$$ (1.9)

The equilibrium constant, K, for the solution reaction is given by Eq. (1.10).

$$E_P^0 - E_M^0 = RT \ln K.$$ (1.10)

E_P^0 values are determined from plots of charge *vs.* change in, for example, the optical absorbance of the protein.

1.2.1.2 Redox Potentials and Stoichiometry of Electron Transfer in Biological Electron Transports

The standard redox potentials of a number of biologically important redox couples are given in Table 1.2 [4]. Systems having a more negative standard redox potential than the H_2/H^+ couple have a greater tendency to lose electrons than hydrogen: those with a more positive potential have a lesser tendency to lose them. The standard redox potentials of various biological redox systems allow us to predict the direction of flow of electrons from one redox couple to another under standard conditions. Furthermore, it is possible to calculate the final equilibrium resulting when electrons flow from one redox system of known standard potential to another of known potential, as well as the free-energy changes occurring during such reactions.

Respiratory Electron Transport Chain. Three major classes of redox enzymes participate in the mainstream of electron transport from organic substrates to molecular oxygen. In the order of their participation these are the NAD(P)-linked dehydrogenases, FAD/FMN-linked dehydrogenases, and cytochromes.

The sequence of electron-transfer reactions in the respiratory chain is consistent with the standard redox potentials of the different electron carriers in that the potentials become more positive as electrons pass from substrate to oxygen (Fig. 1.7). First, NADH donates its two electrons together with one proton (H^+) to NADH dehydrogenase containing FMN. FMN accepts an additional proton from the medium inside the membrane and is thereby reduced to $FMNH_2$. The soluble coenzyme Q(CoQ), also called ubiquinone, serves as a highly mobile carrier of electrons between the flavoproteins and the cytochromes of the electron transport chain. Unlike the other components of the respiratory chain, the quinones probably migrate through the membrane.

The cytochromes of the respiratory chain exist in a complex known as cytochromes b, c_1, c, a, and a_3. During the 1920s, KEILIN carried out pioneering studies on the cytochromes, but he did not at that time realize that the terminal oxidase belongs to this group of respiratory proteins. It was not until 1939 that he discovered that his cytochrome a, having an absorption line close to 600 nm, was heterogeneous and had a component cytochrome a_3. The term cytochrome a was retained for the component that does not react directly with dioxygen. It

Table 1.2. Redox potential of biologically important redox couples

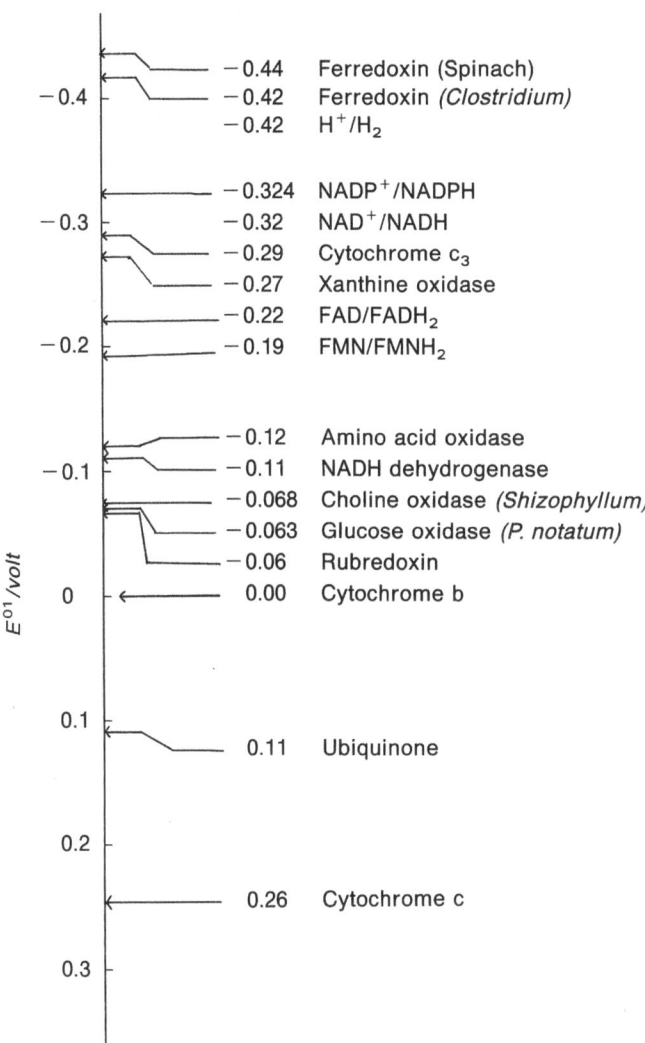

−0.44	Ferredoxin (Spinach)
−0.42	Ferredoxin *(Clostridium)*
−0.42	H^+/H_2
−0.324	$NADP^+/NADPH$
−0.32	$NAD^+/NADH$
−0.29	Cytochrome c_3
−0.27	Xanthine oxidase
−0.22	$FAD/FADH_2$
−0.19	$FMN/FMNH_2$
−0.12	Amino acid oxidase
−0.11	NADH dehydrogenase
−0.068	Choline oxidase *(Shizophyllum)*
−0.063	Glucose oxidase *(P. notatum)*
−0.06	Rubredoxin
0.00	Cytochrome b
0.11	Ubiquinone
0.26	Cytochrome c

is now firmly established that the minimal functional unit of cytochrome oxidase (terminal oxidase) contains two heme groups in different environments, cytochromes a and a_3, as well as two copper atoms with distinct properties [5].

Although many cytochromes have been highly purified, with one exception they are usually very tightly bound to the mitochondrial membrane and difficult to obtain in soluble and homogeneous form. Cytochrome c is the exception, which is very easily extracted from mitochondria by strong salt solutions. The iron protoporphyrin groups of cytochrome c are covalently linked to the protein via thioether bridges between the prophyrin ring and two cysteine residues in the peptide chain, presumably formed by addition of the –SH group across the double bond of the 2- and 4-vinyl groups of the protoporphyrin. The fifth and sixth

9

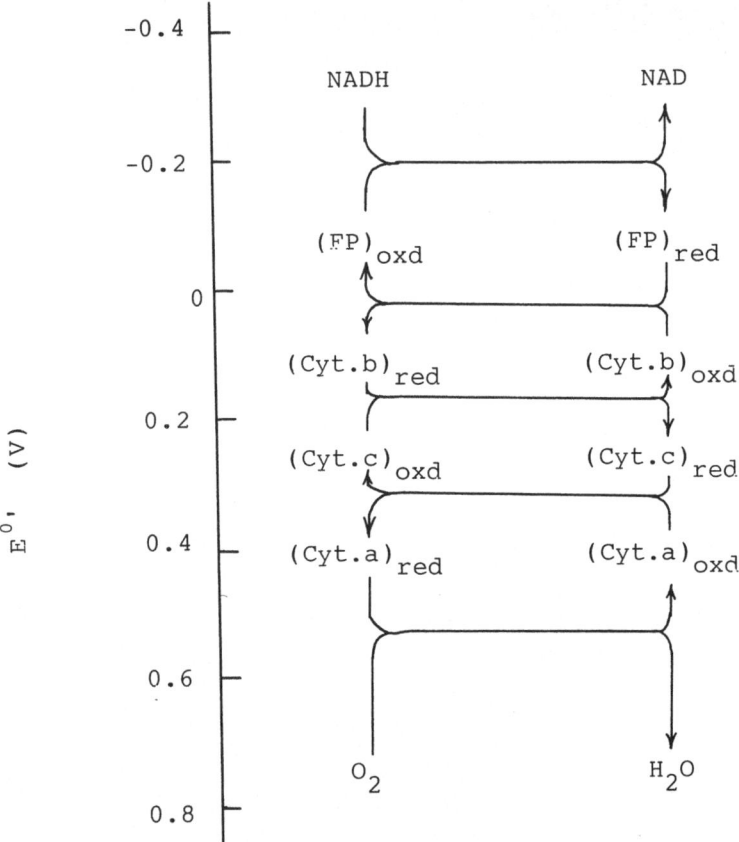

Fig. 1.7. Electron transfer in the respiratory chain

coordination positions of iron are believed to be occupied by a histidine residue and a methionine residue, which prevent cytochrome c from reacting with oxygen of carbon monoxide at pH 7.0. The structure of cytochrome c is illustrated in Fig. 1.8 [6]. The redox potential, $E^{0'}$, of cytochrome c is determined as 0.01 V.

While cytochrome oxidase contains two distinct cytochromes, only one type of heme, designated heme a, can be isolated from it. Thus the different properties of cytochrome a and a_3 must be derived from differences in the binding to the protein part of the enzyme. The iron atom of cytochrome a is in a low-spin state. Cytochrome a_3, on the other hand, has its iron atom in a high-spin state. The structure of heme a, as isolated by extraction, is given in Fig. 1.9.

The redox properties of cytochrome oxidase were monitored by optical measurements. Measurements at 830 nm yields a straight line, with n in the Nernst equation equals to 1. This finding indicates that the 830 nm absorbance, which is mainly due to Cu_A^{2+}, results from the oxidation-reduction of a one-electron center.

10

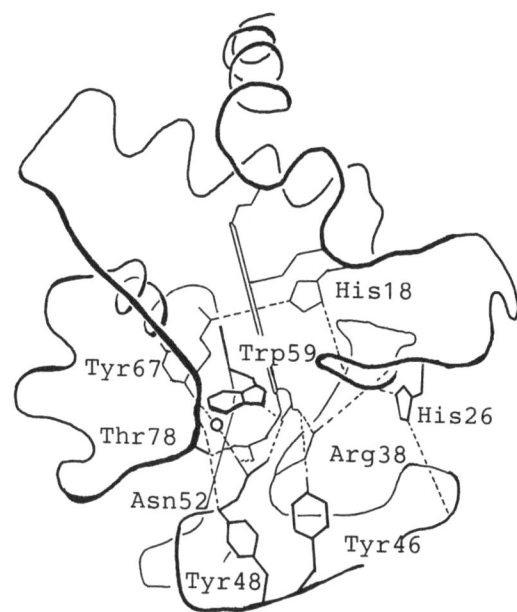

Fig. 1.8. Structure of cytochrome C

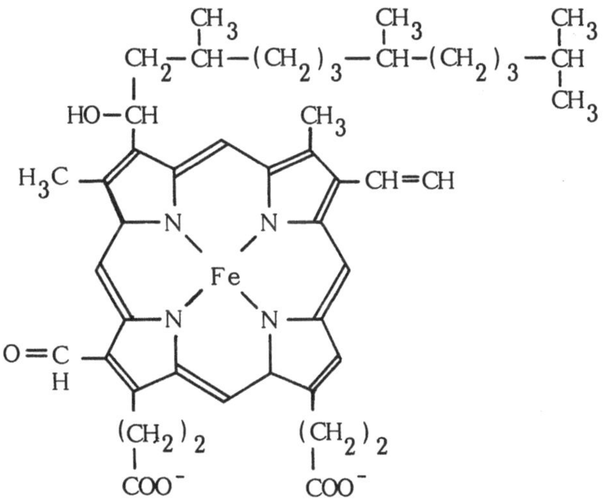

Fig. 1.9. Structure of heme a

Photosynthetic Electron Transport. All oxygen-evolving photosynthetic cells contain both photosystems I and II, whereas photosynthetic bacteria, which do not evolve oxygen, contain only one photosystem and have their own characteristic assembly of light-absorbing pigments.

Photosystem I in higher plants and most algae chloroplasts is excited by far-red light (700 nm), while photosystem II depends upon higher energy red light (680 nm). The P700 and P680 reaction centers are contained in photosystem I and

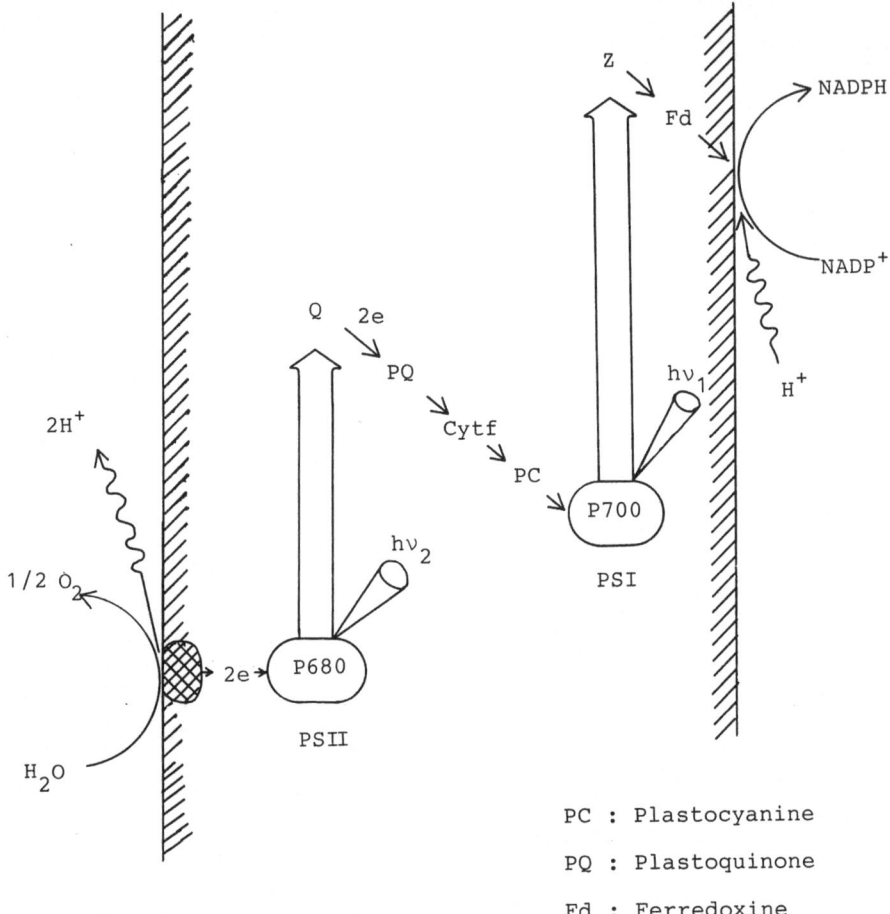

PC : Plastocyanine

PQ : Plastoquinone

Fd : Ferredoxine

Fig. 1.10. Electron-transfer chain in the photosynthetic process in the thylakoid membrane

II, respectively. They contain one or two chlorophyll molecules capable of producing a separation of positive and negative charges across the thylakoid membrane that will initiate electron transport, while the whole photosynthetic unit consists of 200–400 antenna chlorophyll molecules.

The entire electron-transfer chain is schematically illustrated in Fig. 1.10. Electrons and protons generate from the H_2O/O_2 couple which has a strong positive redox potential ($E^{0\prime} = +0.811$ V), while the $NADPH/NADP^+$ couple possesses a negative value of $E^{0\prime} = -0.322$ V. An electron transport against a potential gradient is driven by light.

The carrier sequence in the photosynthetic electron transport between photosystem II and I has been elucidated by many of the same methods used in analysis of the respiratory electron transport, by piecing together information on the standard redox potentials of the carriers, by studying their behavior spectroscopically, and by the action of inhibitors and artificial electron carriers.

12

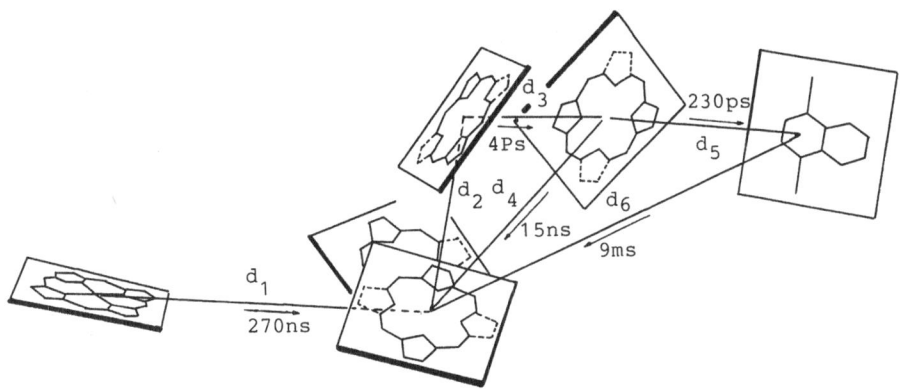

Fig. 1.11. Rate of electron transfer in the reaction center of the bacterial photosystem

An important electron carrier of photosynthesis is ferredoxin. Ferredoxins are found in all photosynthetic cells, but only in certain non-photosynthetic obligate anaerobes. Spinach ferredoxin has a molecular weight of 11 600, containing two iron atoms which are bound to two specific sulfur atoms. Although, ferredoxin can be reduced and reoxidized *via* the one-electron steps, it is not certain whether this necessarily occurs *via* Fe(III) to Fe(II).

The redox potentials, measured at different partial pressures of H_2 and at different pH values, were found to be: $E^{0'} = -0.42$ V for spinach ferredoxin and $E^{0'} = -0.39$ V for *clostridial (Clostridium pasteurianum)* ferredoxin. The redox potentials of spinach and clostridial ferredoxin were found to involve the transfer of one electron per molecule [7].

In contrast to plant photosynthesis, photosynthetic bacteria contain only one photosystem, which is not associated with oxygen generation. Since the photosynthetic reaction center can easily be isolated from photosynthetic bacteria, elucidation of the mechanism has intensively been progressed with the bacterial photosystem. The reaction center consists of three polypeptide chains, which contains four molecules of bacteriochlorophyll (BChl), two molecules of bacteriopheophytin (BPheo), one molecule of nonheme, and two molecules of quinone. Picosecond spectroscopy has revealed that the photoexcited singlet of the reaction center P*, transfers an electron to the adjacent BChl-a in the reaction center B, within 1 ps. The rate of electron transfer in the photoexcited charge separation was determined as shown in Fig. 1.11 [8].

Nitrogen Fixation. The fixation of nitrogen requires the cooperative action of the host plant and bacteria present in the root nodules; i.e., symbiotic nitrogen fixation. The mechanism of nitrogen fixation has long been a challenging biochemical problem.

Nitrogen reduction is catalyzed by nitrogenase. Nitrogenase, isolated from *C. pasteurianum* has a two component system, consisting of a large protein (MW = 220 000 with the subunits $2 \times 50000 + 2 \times 50000$) containing 2 Mo, 20 Fe, and 20 S, possibly in the form of iron-sulfur clusters (molybdo-ferredoxin), and a smaller protein (MW $= 55\,000$) containing iron-sulfur but not molybdenum (azo-ferredoxin).

The nitrogenase system catalyzes the six-electron reduction of nitrogen to ammonia in three successive two-electron steps where only electrons are transported. The following mechanism has been proposed [9]:

$$E + 2e^- + 2H^+ \xrightarrow[\;]{\text{2ATP} \quad \text{2ADP}} E^* \cdot 2H, \tag{1.11}$$

$$E \cdot 2H + N_2 \rightleftharpoons E^*N_2 + H_2, \tag{1.12}$$

$$E^*N_2 + 2e^- + 2H^+ \xrightarrow[\;]{\text{2ATP} \quad \text{2ADP}} E^*N_2H_2, \tag{1.13}$$

$$E^*N_2H_2 + 2e^- + 2H^+ \xrightarrow[\;]{\text{2ATP} \quad \text{2ADP}} E^*N_2H_4, \tag{1.14}$$

$$E^*N_2H_4 + 2H^+ \longrightarrow 2NH_3 + E. \tag{1.15}$$

1.2.2 Electrochemical Reactions of Enzymes and Coenzymes

1.2.2.1 NAD(P)$^+$ and NAD(P)H

Cathodic Reduction of NAD(P)$^+$. The electrochemical reduction of NAD(P)$^+$ has extensively been studied by voltammetry and polarography [10–17]. NAD$^+$ gives a two-step reduction wave at a mercury electrode in the absence of molecular oxygen. The electrochemically generated NAD radicals are dimerized at a potential of -1.2 V *vs.* SCE (saturated calomel electrode), although the conversion proceeds very slowly (Fig. 1.12).

$$NAD^+ + e^- \xrightarrow[\text{ca.} -1.2\,\text{V}]{\text{Electrode}} NAD^{\boldsymbol{\cdot}} \xrightarrow[\text{ca.} -1.58\,\text{V}]{H^+, e^-} NADH, \tag{1.16}$$

$$NAD^+ + 2e^- \xrightarrow{\text{Enzyme}} NADH. \tag{1.17}$$

Free NADP$^+$ gives also a similar two-step reduction wave in an alkaline solution above pH 9 [18]. Tetraalkyl ammonium salts and carbonate buffers are commonly used to obtain these waves. At pH 9.3, NADP$^+$ in a carbonate buffer containing 0.4 M tetraalkyl ammonium bromide gives a two-step wave at a dropping mercury electrode [19–21]. The effectiveness of a supporting electrolyte such as tetraethyl ammonium chloride for the depression of dimerization also has been shown by SCHMAKEL et al. [21, 22], who studied the rate constant and activation process; the products were 1,6-reduced NADH besides the 1,4-reduced form (Fig. 1.13). Further details of the dimerization in the initial one-electron electrochemical reduction of NAD$^+$ have been studied by BRESNAHAN and ELVING [23]. In the presence of electrolyte in low concentrations, NAD$^+$ adsorbs on the electrode surface at potentials more positive than ca. -0.9 V *vs.* SCE where the one-electron reduction proceeds. The electrochemically produced dimer (NAD)$_2$ may adsorb at -1.20 and -1.32 V in 0.06 and 0.4 M KCl solutions, respectively.

14

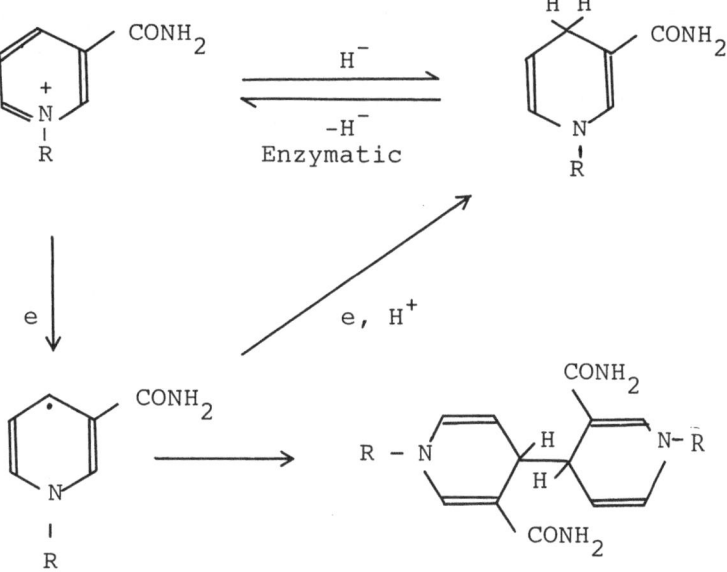

Fig. 1.12. Electrochemical and enzymatical reductions of NAD$^+$

1,4-NADH   1,6-NADH

Fig. 1.13. Reduced forms of NAD   R : Adenosine diphosphoribosyl

NAD$^+$ undergoes both diffusion and adsorption-controlled reduction at low electrolyte concentrations (from 0.06 to 0.1 M). The adsorbed layer of NAD $\cdot$ and/or (NAD)$_2$ is formed upon reduction of dissolved NAD$^+$. The working model for the orientation of adsorbed NAD$^+$ at the electrode/solution interface assumes the adenine ring to be located flat on the surface with the coenzyme in a folded conformation. The NAD radical generates hydrogen peroxide in the presence of molecular oxygen [24]. The dimer is easily oxidized and the product is enzymatically shown to function as a coenzyme. The dimer photochemically dissociates into NAD$^+$ in the absence of molecular oxygen [25]. Therefore one can expect that NAD$^+$ is reduced to NADH without dimer formation under the irradiation of ultraviolet light (254 nm).

On the mercury electrode controlled at -1.1 V $vs.$ SCE, 90% of NAD is reduced to its dimers; 4,4'-dimer (90%) and 6,6-dimer (10%). At -1.8 V reduction,

1,4-NADH (50%), 1,6-NADH (30%), and dimers (20%) are produced [26]. The rate constant of NAD dimerization is $3 \times 10^{-6} \, M^{-1} \cdot s^{-1}$ [22]. Under UV and visible light irradiation the dimer dismutates to NAD^+ and 1,4-NADH [27].

Many an approach has been made to retard the dimerization during the reduction process of NAD^+. AIZAWA et al. have shown that polymer-bound $NAD(P)^+$ is electrochemically reduced to NAD(P)H without the loss of coenzyme activity [19, 28, 29]. $NAD(P)H^+$ was covalently bound to watersoluble alginate as a pendant, which resulted in a significant retardation of intermolecular contact of NAD(P). Electrochemically generated NAD(P) radicals had a very low probability to form dimers. Liquid crystals are also effective on the depression of NAD(P) dimerization in the electrochemical process. On a mercury electrode covered with a thin cholesterol oleate membrane (0.6 μm thick), NAD(P) gives a one-step reduction wave on the polarogram. Potential-controlled electrolysis was carried out with a catholyte (50 ml) containing 0.1 mM NAD^+ and 0.25 M $MgCl_2$ using the liquid crystal membrane electrode (10 cm^2) at a potential of -1.50 V $vs.$ SCE. The yield of NADH was more than 95% when the electrolysis proceeded for longer than 6 h [30]. Under these conditions $NAD(P)^+$ is electrochemically reduced to its radical. The liquid crystal membrane may inhibit the intermolecular association.

NADH is further reduced to its hydrated form by square wave voltammetry under the following conditions: initial potential -5.5 V, final potential -1.40 V, step height 5 mV, frequency 30 Hz, square-wave amplitude 25 mV, and delay time 10 s [31, 32]. This investigation is of interest, since NADH is cathodically detectable.

A large overvoltage is required to reduce NAD(P) to NAD(P)H. However, the large overvoltage may invite severe interference from other reactions on the electrode. To avoid these problems, electron mediators are commonly used to reduce $NAD(P)^+$. Effective conversion of $NAD(P)^+$ to NAD(P)H is performed with the use of electron mediators such as methylviologen [33, 34] and bipyridine-rhodium complex (Table 1.3) [24].

Anodic Oxidation of NAD(P)H. Extensive investigations have been conducted on the anodic behavior of NAD(P)H [35–38]. The anodic oxidation of NAD(P)H can yield 100% enzymatically active $NAD(P)^+$. Less attention has been paid to the oxidation of NAD(P)H than to the reduction of $NAD(P)^+$, although this has been mentioned in a few reports [15, 18].

Voltammetric studies on the oxidation of NAD(P)H show that the adsorption of pyridine nucleotide is indeed significant especially in the case of a platinum electrode in 0.1 M pyrophosphate buffer (pH 7.6) [39, 40]. The degeneration of the platinum anode is significantly decreased by working at pH 8.1 [41].

Only one anodic wave appears in the current-potential curve for NADH oxidation, in contrast to the cathodic polarogram of NAD^+ where a two-step wave is observed. This suggests that the oxidation of NADH proceeds in one step by a two-electron reaction [39].

The direct electrochemical oxidation of NADH by the platinum electrode implies a high overvoltage [41–45]. Oxidative pretreatment of the carbon electrodes results in reduction of the over-potential by ca. 200–250 mV [42]. Several groups

16

Table 1.3. Mediator-aided reduction of NAD(P)$^+$

Electrode	Electron transfer system	Controlled potential (V vs. SCE)
Tungsten	$*$ Ferredoxin-NADP reductase $*$ Lipoamide dehydrogenase	-0.7
Tungsten	$*$ Lipoamide dehydrogenase DTT : Dithiothreitol	-1.0
Graphite		

have investigated the use of mediators for the NADH oxidation at low applied potentials [46–51]. When o-quinones are employed as mediators, the oxidative peak potential can be reduced from 420 to 250 mV $vs.$ Ag/AgCl electrode in cyclic voltammetry investigations [48]. HUCK and SCHMIDT [47] have shown that chloranil (tetrachloro-p-benzoquinone) is an effective catalyst for the electrochemical oxidation of NADH to NAD$^+$.

3-β-Naphthoyl Nile Blue as a catalyst for NADH oxidation in combination with hydrogenases has been shown suitable for the determination of the corresponding substrates by a graphite electrode [52]. Phenazine methosulfate (PMS$^+$) can also be used as a hydrogen carrier of NADH with a Clark-type oxygen electrode [52]. However, the enzymes are easily denatured by PMS$^+$ [54]. ALBERY and BARTLETT [53] have shown that conducting organic salt (NMP$^+$TCNQ$^+$) electrodes readily oxidize NADH; however these organic conductor electrodes degenerate after a few days (Table 1.4).

17

Table 1.4. Electrocatalytic electrode systems for oxidation of NADH

Electrode	Mediator
Graphite	Chemically modified with Meldola Blue (7-dimethylamino-1,2-benzo-phenoxazine)
	Impregnated with O-quinones
	Impregnated with chloranil
	Coupled with 3-β-Naphthoyl Nile Blue and dehydrogenase
	Coupled with phenazine methosulphate (PMS) and dehydrogenase
Carbon	Oxidatively pretreated
NMP$^+$TCNQ$^+$	

1.2.2.2 FAD and FMN

The electrochemical reduction of flavin adenine dinucleotide (FAD), flavin mononucleotide (FMN), and riboflavin (RF) is characterized by two superimposed reversible one-electron processes. The one-electron reduction of flavins yields a radical anion (semiquinone) with the highest spin density being located

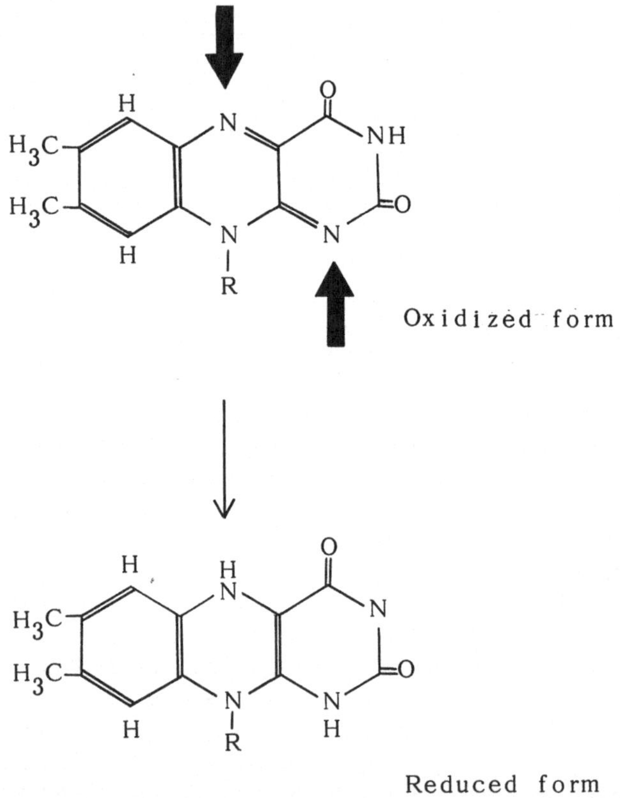

Oxidized form

Reduced form

Fig. 1.14. Semiquinone forms of FMN and FAD

18

at the N (5) position (Fig. 1.14). The product of the two-electron reduction is 1,5-dihydroflavin.

Flavin nucleotides are the prosthetic groups of flavoenzymes such as glucose oxidase and xanthine oxidase. Flavin nucleotides strongly bind to apoproteins, which causes steric hindrance to electron transfer with the electrode.

1.2.2.3 Flavoenzymes

Flavoenzymes appear not to be reduced at the electrode.

Xanthine Oxidase. Xanthine oxidase has a molecular weight of 270 000, containing two FAD groups, eight FeS, and two Mo. Although the prosthetic groups of xanthine oxidase are not polarographically active at the DME, phase-selective AC polarograms of the enzyme exhibit peaks at -0.430, -1.590, and -0.680 V *vs.* SCE [55]. Free FAD gives a peak at -0.430 V under the same solution conditions. The formal potential for this redox process was determined to be -0.534 V *vs.* SCE.

Glucose Oxidase. Glucose oxidase (GOD) from *Penicillium notatum* has a molecular weight of 186 000. It consists of two identical subunits, each containing one FAD group. During chromatographic purification, protease impurities cause a partial degradation of GOD leading to a decrease in molecular weight of the enzyme molecule. Nevertheless, the remaining fragments possess a high specific activity (200 IU mg^{-1}), as proved by the oxygen consumption method.

Such GOD fragments give a well-shaped peak in differential-pulse polarograms, whose peak potential at pH 7 is -0.340 V *vs.* SCE [56]. This value corresponds very well to the redox potential of -0.305 V *vs.* SCE measured at pH 5.3. In the same solution, free FAD shows a peak with a potential at -0.440 V. GOD is reduced at the mercury electrode at more positive potentials than that of the free prosthetic group. The GOD peak is attributed not to dissociated FAD but to the apoprotein-bound FAD. This conclusion is supported by the following facts: The peak height increases and the peak potential coincides with that of free FAD in 4 *M* guanidine hydrochloride solution (pH 6.8).

The peak height in the differential-pulse polarographic measurement is proportional to the GOD concentration, while the peak potential remains constant. The diffusion coefficient is estimated as 5×10^{-7} cm$^{-2} \cdot$s^{-1} from the Ilkovic equation. The limiting current for reduction of GOD corresponds roughly to 30% of the value calculated.

A DC polarographic wave of GOD is shown at a half-wave potential of -0.38 V *vs.* SCE. The slope of $dE_{DME}/d\{\log[i/(i_d-i)]\}$ plots was 55 mV. The enzymatically reduced GOD gives a polarographic oxidation wave. The height of this wave is the same as the reduction wave of the nonreduced species. The half-wave potential is shifted by some 50 mV in a positive direction. The enzymatically reduced GOD is finally reoxidized at the DME.

These results suggests that the oxidation–reduction process of GOD at the electrode should be reversible.

Table 1.5. Formation constants of semi-quinone and electron transfer rate constants for ChOD and COD

	ChOD	COD
K_{sem}	1.5 (pH 7.4)	1.5 (pH 0.8)
K'_s (s^{-1})	1.6×10^3 (pH 5.4)	0.6×10^3 (pH 8.9)

K_{sem}: Formation constant of semiquinone.
K'_s: Rate constant of electron transfer.

In contrast to the reversible electrode reaction of the GOD fragments from *Penicillium notatum*, GOD from *Aspergillum niger* does not exhibit a reduction peak at platinum or glassy carbon electrodes. However, the electrode reaction of intact GOD both from *Penicillium notatum* and *Aspergillum niger* proceeded at a significant rate at a modified gold electrode, as demonstrated by spectroelectrochemical methods.

Azoflavin and Amino Acid Oxidase. The FMN containing azoflavin form *Azotobacter vinelandii* exhibits a peak at -0.40 V *vs.* SCE in phase-selective AC polarograms. KUZNETSOV et al. attributed this peak to the reduction of the free prosthetic group [57]. Similarly, amino acid oxidase gives a peak at -0.36 V *vs.* SCE [55].

Cholesterol Oxidase and Choline Oxidase. Cholesterol oxidase (ChOD) from *Shizophyllum commune* has a molecular weight of 53 000 and contains one FAD group. Choline oxidase (COD) from *Cylindrocarpon didymum M-1* has a molecular weight of 120 000, containing two FAD groups. The prosthetic groups of these enzymes are covalently bound to the corresponding apoprotein. Therefore, it should be less possible that the prosthetic FAD group dissociates from the apoprotein.

Both ChOD and COD adsorb so strongly on the mercury electrode surface that the adsorbed enzymes give a reversible electron exchange between the prosthetic group FAD and the HMDE [58, 59]. It was estimated from cyclic voltammetry of the adsorbed protein that Γ^{max} of ChOD should be 4×10^{-12} mol·cm^{-2}. The Γ^{max} of COD was comparable to that of ChOD. The peak potentials of ChOD and COD in cyclic voltammograms were -0.31 (pH 7.0) and -0.33 (pH 7.0) V, respectively. Both DC and AC cyclic voltammograms were analyzed according to the theoretical equation of the two-step surface redox reaction. The formation constant of semiquinone, K_{sem}, and the apparent rate constant of electron transfer, k'_s, were calculated for ChOD and COD as listed in Table 1.5 [60].

1.2.2.4 Heme Proteins

Heme proteins take part in respiration as oxygen carriers (hemoglobin) and in the storage of oxygen (myoglobin). Also, they participate in the reduction of perox-

ides (catalase, peroxidase) and in electron-transfer reactions between dehydrogenases and terminal electron acceptors in the respiratory chain (cytochromes). The
prosthetic group of all these proteins is an iron porphyrin called heme (Fig. 1.9).
Four coordination positions of the heme iron are occupied by pyrrole nitrogens
of the porphyrin ring. Two additional bonds can be formed by the iron, one on
either side of the heme plane (fifth and sixth coordination position).

Cytochrome b. The prosthetic group of cytochrome b is iron porphyrin IX, which
is not covalently bound to the protein *via* heme substituents, i.e., the same mode
as for Hb and Mb. Typical b-type cytochromes do not react with oxygen, CO,
or cyanide.

Cytochrome b_5 presumably plays a role in delivering electrons from NADH
via cytochrome b_5 reductase to cytochrome P-450 or to a fatty acid desaturation
system. Cytochrome P-450, which catalyzes the hydroxylation of drugs and ste-
roids, also belongs to the b-type cytochromes.

Solubilized fractions of liver cytochrome P-450 exhibit a polarographic reduction wave having $E_{1/2} = -0.580$ V *vs.* SCE due to the electrochemical reduction
of ferri P-450 [61]. Chromatographically purified P-450, however, does not exhibit any polarographic wave, which indicates that the reduction of cytochrome
b_5, a constituent of the microsomal fractions, may take place at the polarographic
wave of -580 mV. Nevertheless, during the large-scale controlled-potential electrolysis of solubilized P-450 at a potential of -0.80 V *vs.* SCE at a mercury pool
electrode, a product containing ferrous P-450 was obtained. At a carbon cathode
or vibrating nickel electrode, an inactive but not completely destroyed reduction
product of cytochrome P-450 was obtained. It should be noted that cytochrome
P-450 from *Pseudomonas putida* gives a reversible redox wave in the presence of
4,4′-bipyridine at a nickel electrode [62].

Cytochrome c. The heme group of c-type cytochromes is covalently attached to
two cysteine side-chains of the protein (Fig. 1.8). In both the oxidized and reduced
forms of the protein, methionine and histidine are in the fifth and sixth coordination position of the iron. Thus, the heme is nearly inaccessible to the surrounding
solvent, and occupies only about 2% of the molecular surface. In the mitochondrial respiratory chain, it forms a complex with cytochrome oxidase, to which one
electron is transferred. However, it accepts one electron in the reaction with cytochrome c_1. Cytochrome c_1 has the same prosthetic group and mode of attachment to the protein as cytochrome c.

Polarographic studies on cytochrome c have extensively been made by BESTO
[63], SCHELLER [64–66], KUZNETSOV [67, 68], and HALADJIAN [69, 70]. These reveal
that cytochrome c gives an irreversible reduction wave at a half wave potential
of -0.1 V (pH 6.0, 100 μM cytochrome c) which is extremely shifted from its formal potential of 0.01 V. The oxidation wave of reduced cytochrome c does not
appear in the potential range below $+0.1$ V.

Cytochrome c strongly adsorbs on the HMDE surface [67]. The saturated adsorption of cytochrome c is estimated to be 7.9×10^{-12} mol·cm^{-2} by differential
capacitance measurement and tracer measurement (Table 1.6) [66]. Since the experimental data is far less than the estimated saturated adsorption, cyto-

Table 1.6. Electron promotors for cytochrome c

Electron promotor	Electrode
4,4′-Bipyridine	Au, Pt
1,2-Bis(4-pyridyl)ethylene	Au, Pt
Bis(4-pyridyl)sulfide	Au
Bis(4-pyridyl)disulfide	Au

chrome c adsorbs on the HMDE surface presumably in an unfolded form. The electron transfer process of cytochrome c has been interpreted by the models proposed by SCHELLER [65] and KUZNETSOV (Fig. 1.15) [68].

Potential-controlled reduction of cytochrome c at a platinum electrode gives a reduced form of cytochrome c [71]. However, electron transfer of cyto-

Electrode

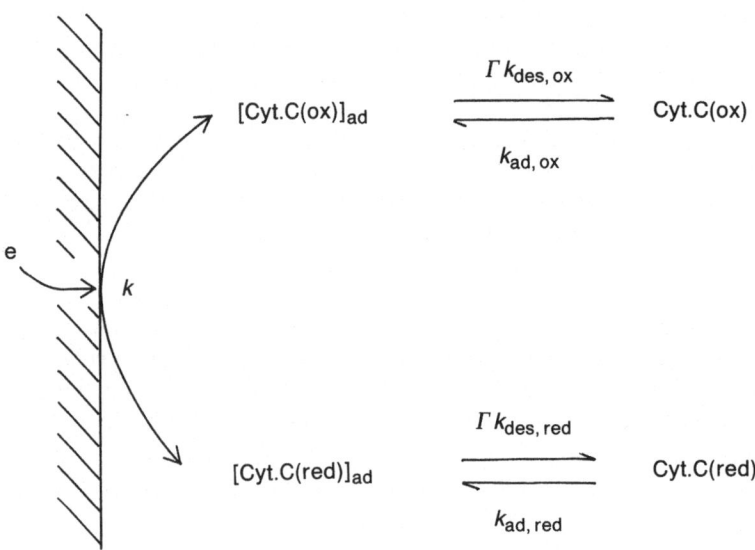

$k_{ad, ox} = 3.3 \times 10^{-2}\, cm \cdot s^{-1}$,　$\Gamma_{des, ox} = 5.9 \times 10^{-9}\, mol \cdot cm^{-2} \cdot s^{-1}$

$k_{ad, red} = 2.5 \times 10^{-2}\, cm \cdot s^{-1}$,　$\Gamma_{des, red} = 6.3 \times 10^{-9}\, mol \cdot cm^{-2} \cdot s^{-1}$

$k_s = 30\, s^{-1}$,　$k = k_s \exp\left[(-nF/RT)(E - E^0)\right]$

Fig. 1.15. Rate constants of electrode process of cytochrome C

chrome c at a platinum electrode is very slow. In contrast, cytochrome c gives distinct redox waves in cyclic voltammograms at a tin-doped In_2O_3 electrode [72].

The effects of electron promoters on the electron transfer of cytochrome c at the electrode interface have been intensively studied. In the presence of 4,4'-bipyridine, cytochrome c gives a sharp cyclic voltammogram at gold and platinum electrodes [73, 74]. Several electron promoters for cytochrome c have been reported, including 1,2-bis(4-pyridyl)ethylene, bis(4-pyridyl)sulfide, and bis(4-pyridyl)disulfide (Table 1.6) [73, 75].

The peak current i_p of cytochrome c is proportional to the concentration of 4,4'-bipyridine and the square root of the potential sweep rate which indicates that the electrode process is reversible. The diffusion coefficient of cytochrome c is estimated from i_p to be 9.4×10^{-7} $cm^2 \cdot s^{-1}$ at a gold electrode and 6.0×10^{-7} $cm^2 \cdot s^{-1}$ at a platinum electrode. The half wave potential, $(E_{pc} + E_{pa})/2$, coincides with the formal potential $E^{0\prime}$ of cytochrome c. These characteristics suggest that the reversible redox reaction of cytochrome c takes place in the presence of an electron promoter at gold and platinum electrodes.

The states of adsorbed cytochrome c have been studied by surface-enhanced raman spectroscopy (SERS) [76].

Cytochrome c_3. The multiheme protein cytochrome c_3, an electron-transfer protein from the sulfate-reducing bacteria *Desulfovibrio vulgaris* (strain *Miyazaki*), was the first example of a heme protein exhibiting a reversible electrode reaction at a mercury electrode [77–80]. Cytochrome c_3 from *D. vulgaris* (strain *Hildenborough*) and *D. desulfuricans* (strain *Norway*) were found to give a reversible redox wave at a mercury electrode [81–84]. Potentiometric titration yields a Nernst slope of 90 mV and an apparent redox potential, $E^{0\prime}$, which on the average for four different hemes, is -0.528 V *vs.* SCE.

At the DME, ferricytochrome c_3 gives a single, well-defined reduction wave. The half-wave potential is -0.527 V *vs.* SCE. Polarograms for a solution containing both the ferric and ferrous forms were symmetrical about the half-wave potential, which agrees very well with the $E^{0\prime}$ value of -0.527 V. The diffusion coefficient of ferricytochrome c_3, determined from pulse polarographic data, is in accordance with that determined by hydrodynamic methods. The electron-transfer reaction rate is too fast to be determined from current *vs.* time curves with a time resolution of the pulse polarography of 20–80 ms. The fast heterogeneous electron transfer is also confirmed by results of cyclic voltammetry – i.e., the ratio of the reduction to oxidation peaks was almost unity and a plot of the peak current versus the square root of the voltage sweep rate was linear. The height of the peak current was only 1/2.5 of the theoretical value of one electron exchange reaction of ferric and ferrous cytochrome c_3. As a reversible and controlled mass transport of cytochrome c_3 molecules from the bulk solution is very unlikely, electron transfer probably takes place through the first adsorption layer.

Disagreement about the number of electrons transferred per molecule determined by potentiometry and cyclic voltammetry has been resolved using a mode employing four reversible redox reactions with separate but closely spaced standard potentials. The best fit values for the individual redox potentials from experimental cyclic voltammetry and differential pulse polarography data are:

$$E_1^{0'} = -0.467 \text{ V } vs. \text{ SCE} \qquad E_2^{0'} = -0.513 \text{ V } vs. \text{ SCE}$$

$$E_3^{0'} = -0.539 \text{ V } vs. \text{ SCE} \qquad E_4^{0'} = -0.580 \text{ V } vs. \text{ SCE}.$$

An alternative suggestion is that all the hemes are identical and interacting. The concept of identical hemes implies interactions between the heme residues, but such a model is unable to explain the asymmetry observed in differential-pulse polarography.

Two reduction waves have been observed in differential-pulse polarograms of cytochrome c_3 from *Desulfovibrio vulgaris* and *Desulfovibrio desulfuricans*. The first wave may have been caused by traces of oxygen. BIANCO and HALADJIAN concluded from polarographic and large-scale electrolysis studies that the electron exchange with the electrode is rapid. They found that the protein undergoes a partial degradation during electrolysis.

Cytochrome a. The a-type cytochromes have a different iron-porphyrin prosthetic group called heme a. It differs from the heme in c-type cytochromes in having a formyl group in place of a methyl group at position 8 and a hydrocarbon isoprenoid side-chain at position 2 in place of one of the vinyl groups.

Cytochrome oxidase is the terminal member of the cytochrome chain and the only member able to reduce oxygen to two molecules of H_2O, i.e., four electrons are required for this reduction. Cytochrome oxidase of mammalian mitochondria contains two hemes plus two atoms of copper per functional unit. It has been proposed that the intact cytochrome a_3 and Cu^+ ion can react directly with oxygen whereas cytochrome a cannot. In the next step, oxygen is probably reduced in a two-electron process to a peroxide structure and subsequently to two water molecules.

1.2.2.5 Iron-Sulfur Protein

Iron-sulfur proteins (ISP) contain nonheme iron bound to sulfur. The three types of structurally related iron-sulfur centers are illustrated in Fig. 1.16. Many proteins contain iron-sulfur centers in association with other prosthetic groups.

Ferredoxin (Fd) from *Clostridium pasteurianum* gives a DC polarographic wave at the DME [85]. Electrode reactions of Fd from various sources have been investigated by DC, AC, differential polarography, and voltammetry [86–90]. The reduction wave, however, shows a reflection of adsorbed ferredoxin, which might be denatured on the electrode surface.

The simplest iron-sulfur proteins are the rubredoxins, which contain a single iron in tetrahedral coordination with the sulfhydryl groups of four cysteines of the single polypeptide chain ($MW = 6000$, $E^{0'} = -0.297 \text{ V } vs. \text{ SCE}$).

The character of the electrode process with FeS proteins is determined both by the conformational stability of the protein and by the time of exposure of the protein to the electrode. At the DME, a polarographic adsorption prewave is observed with several FeS proteins. The native cluster may be involved in this pro-

24

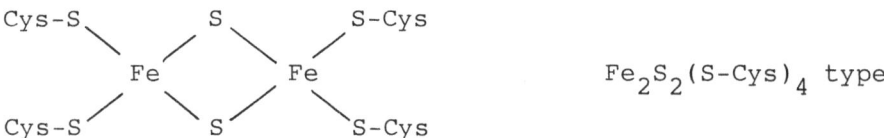

Fe(S-Cys)$_4$ type

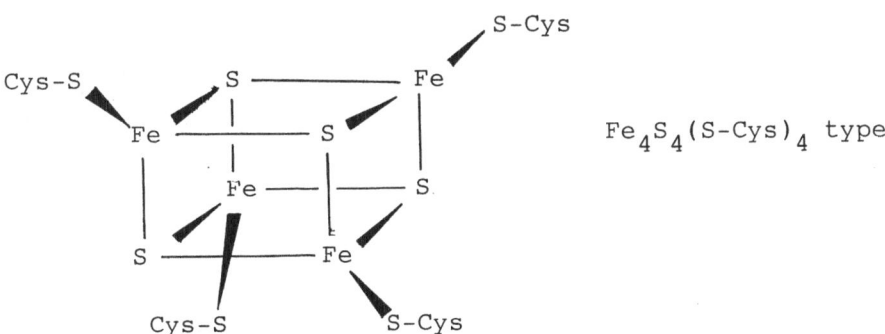

Fe$_2$S$_2$(S-Cys)$_4$ type

Fe$_4$S$_4$(S-Cys)$_4$ type

Fig. 1.16. Three types of iron-sulfur centers

cess, since this wave is absent in polarograms of the apoprotein. Obviously, the confirmation of the halobacterial ferredoxin is not drastically influenced by adsorption at the DME. At HMDE or during large-scale controlled-potential electrolysis – the FeS centers may be disrupted, with resultant denaturation of the protein. A summary of electrochemical data on several proteins is presented in Table 1.7.

Table 1.7. Saturated adsorption Γ^{max} of proteins at the mercury electrode surface

Protein	Γ^{max} (mol cm^{-2})
Cytochrome C	7.9×10^{-12}
Cytochrome C$_3$	9.2×10^{-12}
ChOD	4×10^{-12}
COD	4×10^{-12}

1.2.2.6 Disulfide Bonds in Peptides and Proteins [4]

The thiol-disulfide redox reaction is best illustrated in vivo by the tripeptide glutathione (GSH) (γ-glutamylcysteinylglycine), which acts as an intracellular redox pool. The steady-state ratio GSH/GSSG (GSSG: oxidized form of GSH) is typically about 20. The disulfide moieties of a few proteins are also involved in redox reactions. For example, thioredoxin (MW = 12 000) catalyzes the reduction of ribonucleotides to deoxyribonucleotides. Its redox active group contains the structure:

$$\begin{array}{ccccc}
\text{Trp-Cys-Gly-Cys-Lys} \\
\mid \qquad\quad \mid \\
\text{S}\text{———}\text{S}
\end{array}$$

In addition to the SH/SS redox system, the dehydrogenases of lipoamide, glutathione, and thioredoxin contain a flavin group (FAD) in each subunit. The redox potential, $E^{0\prime}$, of the disulfide group in lipoamide dehydrogenase has been determined to be -0.522 V vs. SCE. The respective dehydrogenases transfer electrons from nicotinamide nucleotides to the oxidized glutathione and thioredoxin, while lipoamide dehydrogenase reoxidized the two thiol groups of lipoamide to the disulfide form.

Formation of SS-bridges between polypeptide chains is an inherent step in protein biosynthesis, e.g., the linkage of the A and B chains of insulin. The polarographic behavior of disulfide-containing proteins is strongly influenced by the interaction of the protein sulfur with the electrode metal. The typical strong adsorption leads to blockage of the electrode and may hinder the application of electrochemical techniques to in vivo analyses. However, irreversible protein adsorption leads to modification of the electrode, allowing the coupling of enzymatic with electrochemical reactions. Since then, the reduction and oxidation of the disulfide and sulfide groups of proteins have intensively been investigated by AC and pulse polarography.

Insulin. The peptide hormone insulin (MW = 5730) consists of two peptide chains linked by two disulfide groups, containing an SS-bridge between residues 6 and 11 of chain A. Insulin solutions at pH 1 exhibit a single polarographic reduction wave ($E_{1/2} = -0.25$ V vs. SCE). Like cysteine, insulin exhibits two polarographic reduction waves at pH 7.1. The more negative wave ($e_{1/2} = -0.65$ V vs. SCE) corresponds to reduction of the first adsorption layer. An oxidation wave of the reduced insulin appears at an identical $E_{1/2}$. Both waves merge into a single wave when the temperature is raised to 50 °C or when 15% ethanol is added. The result indicates an acceleration of the exchange rate within the adsorption layer. Thus, the first wave is ascribed to reduction of an RS-Hg$_n$-SR compound formed by reaction between the mercury electrode and disulfide residues. Since the second wave disappears upon denaturation of the protein, this wave may represent reduction of intact SS groups.

In cyclic voltammograms of insulin, two peaks are obtained due to the reduction of the two interchain SS-groups. At -0.60 V vs. SCE, the four-electron re-

duction of an adsorbed monolayer of insulin apparently occurs. At around
-1.20 V vs. SCE, reduction of molecules from the solution takes place.

RNase. RNase (MW = 13 700) is a globular protein containing four SS groups
and consists of a single peptide chain. The polarographic behavior of this compound is very similar to that of insulin. A single reduction wave is observed at
pH 1 and two waves at pH 9.2. Large-scale electrolysis at -2.0 V vs. SCE results
in reduction of all four SS-groups.

Peroxidase. Horseradish peroxidase (HRP), a heme protein, exhibits a reversible
redox reaction in the potential range from -0.4 to -0.7 V vs. SCE at a gold
amalgam electrode. The electrode reaction has been ascribed to reduction of four
SS-groups, since the apoenzyme (where the heme group is removed) gave essentially the same results.

Albumin. Bovine serum albumin (BSA), a globular protein, consists of a single
peptide chain, crosslinked by 17 disulfide bonds. The polarographic reduction of
these SS groups gives rise to a wave similar to that observed for insulin at pH 1.
Cyclic voltammetry of BSA at pH 7.4 at the HMDE showed the presence of an
adsorption peak having a peak potential of -0.76 V vs. SCE under conditions
where the electrode surface is not completely covered with the protein. At a completely covered electrode, the peak potential shifts to -0.63 V vs. SCE. The magnitude of the integrated peak current suggests that at least three or four SS groups
are rapidly reduced during cyclic voltammetry. This behavior indicated a rapid
partial reduction of several SS-bonds followed by a slower reduction initiated by
slow structural changes.

Both oxidized and reduced forms of BSA are adsorbed, with their structure being maintained by intact disulfide bonds. Thus SS-bonds can be reformed upon
oxidation of the SH-groups, as indicated by the single symmetrical oxidation
peak ($E_p = -0.60$ V vs. SCE). This indicates that the two corresponding SH-groups are close enough during the oxidation step to instantaneously form an SS-group. Cyclic voltammograms did not change upon repeated scanning. Thus the
electron transfer process is probably chemically reversible.

Trypsin. Trypsin, like insulin, gives two polarographic reduction waves at low
concentrations at pH 9.2. In cyclic voltammetry and differential-pulse polarography, two overlapping peaks are found.

Immunoglobulin. The pentameric forms of human immunoglobulin (IgM) have
been studied by AC polarography. IgM exhibits two reduction peaks at pH 8.2.
The more negative peak has a kinetic character ($E_p = -0.160$ V vs. SCE), while
the more positive peak ($E_p = 0.22$ V vs. SCE) exhibits the behavior expected for
an adsorption-controlled process. Coulometry reveals that 1.5–2 disulfide bridges
are reduced per IgM molecule. The electrochemical reduction process should
break the inter-subunit bridge and the terminal C-bridge (position 575) between
two large chains.

1.3 Electrochemical Processes Coupled with Enzyme Reactions

1.3.1 Bioelectrocatalysis

The fundamental idea of bioelectrocatalysis consists in the utilization of the exceptional catalytic properties of biologically active substances in electrochemical processes. Bioelectrocatalysis is a unique combination of electrochemical and biochemical reactions, which rests not only on the ability to control at will the oxidizing and reducing ability of the electrode by changing its Fermi level, but on the specificity and selectivity of biochemical catalysis. Of particular value in this context are those enzymes that catalyze oxidations and reductions.

Endeavors have been made to develop bioelectrocatalysis with enzymes that catalyze redox reactions. These redox enzymes often contain, or bind, prosthetic

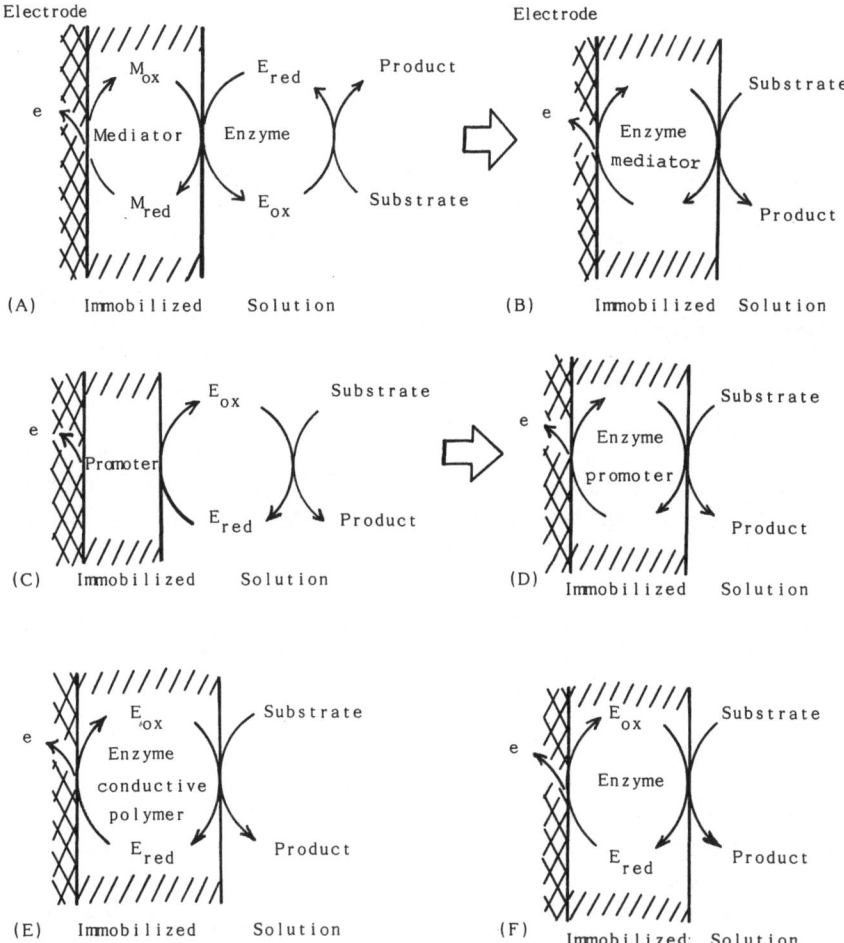

Fig. 1.17. Various bioelectrocatalyses

28

groups that take part in electron transfer. Due to the sophisticated structure of proteins, there is a great difficulty in most cases to achieve electron transfer of enzyme molecules *via* an electrode. Pioneering works of HILL et al. [73, 92], however, have opened the way to realize fast and efficient electron transfer of enzymes at the electrode surface. They modified a gold electrode with 4,4′-bipyridyl, a promoter of electron transfer, not a mediator since it does not take part in electron transfer in the potential region of interest, to accomplish rapid electron transfer of cytochrome c. Their work has triggered intensive investigation of bioelectrocatalysis using modified electrodes.

In the line of research on immobilized enzymes several enzymes have covalently been attached to the electrode surface to enable bioelectrocatalysis. AIZAWA et al. have recently synthesized a conductive enzyme membrane attached to the electrode surface, in which electrons transfer between the enzyme molecules and the electrode [93]. These suggest various approaches for bioelectrocatalysis, which are schematically illustrated in Fig. 1.17.

1.3.1.1 Modified Electrodes
for Direct Electron Transfer of Enzymes

Modified electrodes are fabricated by immobilizing molecules on the electrode surface. MURRAY et al. [94] and MILLER et al. [95] covalently bound functional molecules on the electrode surface in the form of a monomolecular layer. LANE and HUBBARD have modified the electrode surface with strongly adsorbed molecules [96]. Modification with polymer coatings has also proved effective for derivatizing functional electrodes [97]. So far, methods for fabricating modified electrodes may be classified into several categories as follows:

1) Strong adsorption
2) Covalent attachment
3) Polymer coating
   a. Electronic conductive polymer
   b. Ionic conductive polymer.

Comprehensive reviews of modified electrodes have been published recently [98–100].

Charge transfer catalysis has been accomplished by immobilizing electron mediators on the electrode surface. Iron porphyrin covalently bound to a glassy carbon electrode has catalyzed the reduction of molecular oxygen [101]. The overvoltage decreased by about 0.46 V as compared to the bare electrode. The reaction scheme of the charge transfer catalysis is presented in Fig. 1.18.

A covalently modified electrode with face-to-face cobalt porphyrin has provided strong catalytic effects on the reduction of molecular oxygen to water [102]. MURRAY et al. modified a rotating disk electrode with a monolayer of cobalt (*p*-aminophenyl) porphyrin to reduce 1,2-dibromophenylethane to styrene [103].

Charge transfer catalysis also has been achieved by immobilizing a polymer coating on the electrode surface. The polymer-modified electrode shows a greater rate in electrocatalysis as compared to the covalently modified electrode.

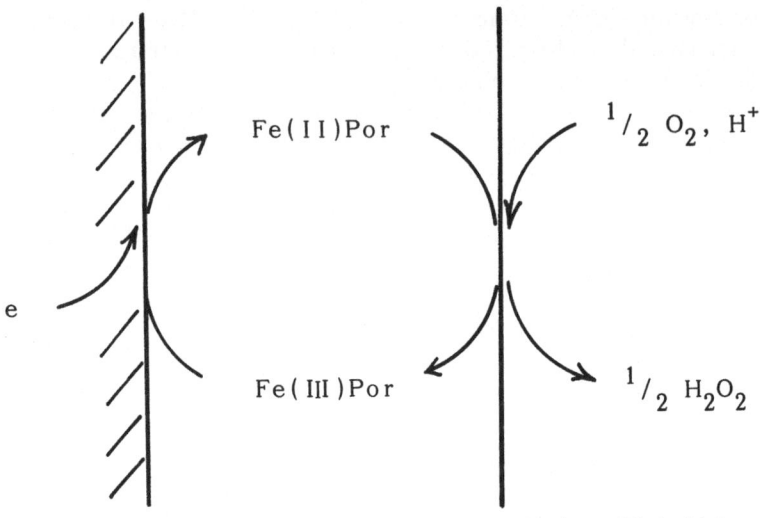

Fig. 1.18. Charge transfer catalysis on an electrode covalently modified with iron porphyrin

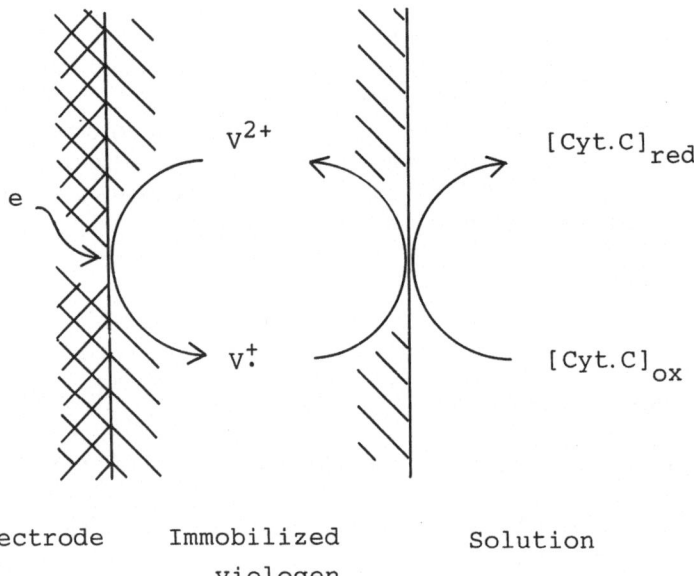

Electrode   Immobilized   Solution

viologen

Fig. 1.19. Reaction scheme of the reduction of cytochrome C at the viologen-modified electrode

Modified electrodes for bioelectrocatalysis use either electron mediators or promoters immobilized on the electrode surface. In both cases, redox enzyme molecules are in solution and in contact with the modified electrodes.

Viologen is one of most common electron mediators for redox enzymes such as cytochrome c and ferredoxin. For this purpose a gold electrode is modified with a thin layer of polymerized viologen. The potential of the gold electrode is controlled at -0.965 V for 5–10 min in a solution containing methylviologen or benzyl viologen. A thin layer of polymerized viologen deposits on the electrode surface. The modified electrode mediates the electron transfer from the electrode phase to redox enzyme molecules in solution.

Cytochrome c was electrochemically reduced at the electrode modified with polymerized viologen [104]. The presumable reaction scheme of the reduction of cytochrome c is illustrated in Fig. 1.19. The electrode donates electrons to immobilized viologen molecules, which is followed by a subsequent electron transfer to cytochrome c in solution. Current-potential curves reflect the electrode reaction of immobilized viologen and the electron transfer from reduced viologen to cytochrome c. A rotating-disk electrode modified with polymerized viologen gave the half-wave potential around the redox potential of viologen ($E^{0'} = -0.5$ V) and the limiting current determined by the diffusion process of cytochrome c. These results strongly support the postulated reaction scheme shown in Fig. 1.19.

It was possible to reduce ferredoxin and myoglobin at an electrode modified with polymerized viologen [105, 106]. The rate constant of electron transfer at the modified electrode was determined as $K'_s = 6.5 \times 10^{-5}$ cm·s^{-1} for spinach ferredoxin [107]. On the other hand, myoglobin at the modified electrode provided a rate constant of electron transfer of $k'_s = 3.88 \times 10^{-11}$ cm·s^{-1}. The electron transfer of ferredoxin was extremely faster than that of myoglobin.

A promoter of electron transfer is not a mediator since it does not take part in electron transfer in the potential region of interest. An electrode modified with promoter molecules has enabled some redox enzymes to directly transfer electrons as shown in Fig. 1.17.

A gold electrode is modified with 4,4'-bipyridyl, a promoter of electron transfer. By most electrochemical criteria, electron transfer of horse heart cytochrome c is fast at the modified electrode; for example, both the D.C. and A.C.-modulated cyclic voltammetries correspond to those of an electrochemically reversible process [73, 92, 108].

The kinetic parameters of cytochrome c were determined by detailed studies at rotating disk and rotating ring-disk electrodes [108]. The rate constants of all the steps in the electrochemical process are presented in Fig. 1.15. The most important feature is that cytochrome c binds to the electrode, and binds productively.

The role of 4,4'-bipyridyl is discussed in relation with the biological system involving cytochrome c. It is suspected that the lysines of cytochrome c surrounding the heme edge bind to carboxylic acid residues on the cytochrome oxidase. In the present case it is possible that they form a hydrogen-bond to the electrode-adsorbed 4,4'-bipyridyl, thus causing the correct orientation. It therefore seems likely that to achieve fast electron transfer at an electrode, the protein should bind tightly.

From analogy to the cytochrome c/cytochrome oxidase coupling, alternative electrodes to the gold-4,4'-bipyridyl system have been exploited by the incorporation of carboxylic acid groups into the graphite surface. Fast electron transfer to cytochrome c results on the negatively charged surface of the electrode.

Thus the synthesis of chemically modified electrodes with 4,4'-bipyridyl and gold, or by covalent attachment of the modifier, should allow for rapid overall

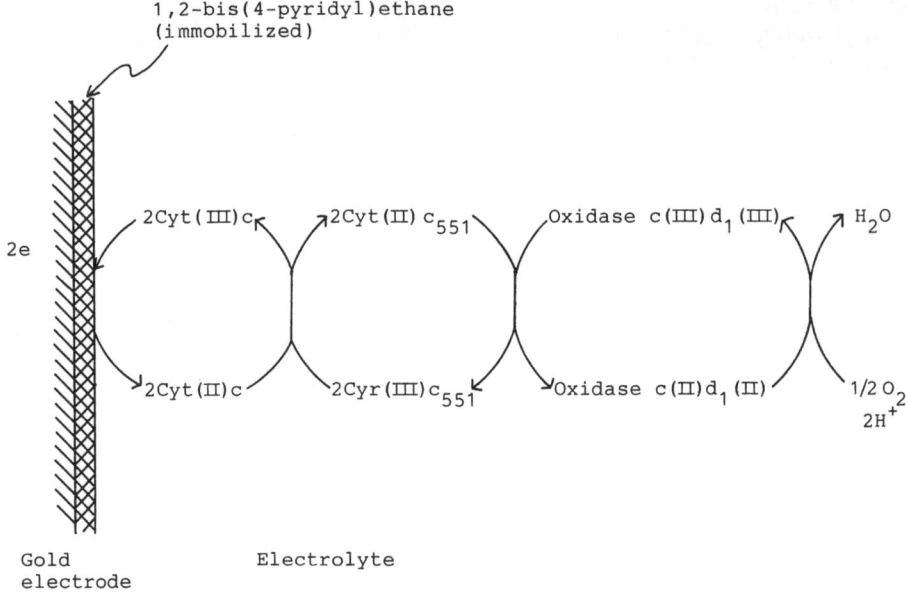

1,2-bis(4-pyridyl)ethane
(immobilized)

2e

2Cyt(III)c ← → 2Cyt(II)c$_{551}$ ← → Oxidase c(III)d$_1$(III) ← → H$_2$O

2Cyt(II)c → 2Cyr(III)c$_{551}$ → Oxidase c(II)d$_1$(II) → 1/2 O$_2$ 2H$^+$

Gold   Electrolyte
electrode

Fig. 1.20. Proposed electron transfer sequence of the reduction of dioxygen to water

electron transfer between enzyme, enzyme complex, or even a more elaborate biological assembly and the electrode.

It has been shown that bis(4-pyridyl)sulfide and bis(4-pyridyl)disulfide are excellent promoters of electron transfer of cytochrome c [75]. Cytochrome c gives a reversible cyclic voltammogram at gold electrodes modified with these promoters.

Bioelectrocatalysis with modified electrodes offer the greatest opportunity for the development of new electrosynthetic methods. Several promising investigations along these lines have been reported.

HILL et al. have coupled the electron transport between a gold electrode, adsorbed with 1,2-bis(4-pyridyl)ethane, and the soluble cytochrome oxidase/nitrite reductase of *Pseudomonas aeruginosa* to efficiently reduce dioxygen to water [109]. The electron transfer sequence has been postulated as shown in Fig. 1.20.

HIGGINS et al. have shown that the electrolysis of a solution containing cytochrome P-450 and the electron transfer proteins putidaredoxin and putidaredoxin reductase, in the cathodic compartment at -0.5 V vs. SCE, gives rise to the formation of an expected product of the enzyme-catalyzed reaction, namely 5-exo-hydroxycamphor [110].

1.3.1.2 Electrode-Bound Enzymes

Several enzymes have been immobilized onto the electrode surface in such a manner as to allow for direct electron transfer between the electrode and immobilized enzyme. Since the mid 1970s, enzyme immobilization techniques have flourished, involving adsorption, covalent, and entrapment methods. A number of enzymes

Table 1.8. Electrode-bound glucose oxidase (GOD)

Base electrode	Immobilization method
Platinum	Adsorption
Carbon paste	Adsorption
Glassy carbon	Adsorption
Graphite	Adsorption
Platinum	Adsorption of allylamide Cross-link with glutaraldehyde Covalent binding of GOD
Octadecyl amine-impregnated carbon paste	Cross-link with glutaraldehyde Covalent binding of GOD
Glassy carbon Graphite	Carboxylation, Hydroxylation Covalent binding of GOD by carbodiimide and cyanulic chloride

have successfully been immobilized onto a variety of solid matrices while still retaining their activity.

Using the adsorption method, cholesterol oxidase (ChOD) from *Shizophyllum commune* and choline oxidase (COD) from *Cylindrocarpon didymum* M-1 were immobilized to the hanging mercury drop electrode (HMDE) surface [58, 59]. ChOD has a molecular weight of 53 000 with one FAD as a prosthetic group, while COD has a molecular weight of 120 000 with two FAD. FAD is covalently bound to the polypeptide chain of ChOD and COD. The immobilized ChOD and COD provided reversible surface redox waves in cyclic voltammetry. The peak potentials were -0.31 and -0.33 V at pH 7.0 for ChOD and COD, respectively. The apparent rate constants of electron transfer, k'_s, were estimated by cyclic voltammetry as 10.6×10^3 s^{-1} at pH 5.4, and 0.6×10^3 s^{-1} at pH 8.9.

Lactate dehydrogenase was adsorbed to the graphite electrode surface [111]. Adsorbed lactate dehydrogenase gave an anodic current proportional to the concentration of lactate in solution. The results indicate a possible direct electron transfer between adsorbed lactate dehydrogenase and graphite.

The covalent method was applied to immobilize xanthine oxidase onto the graphite electrode surface [112]. Glucose oxidase was also immobilized in a similar manner (Table 1.8) [113]. Electrochemical characteristics of these electrode-bound enzymes, however, have not been reported.

Many electrode-bound enzymes have been prepared by the covalent method [114–118]. Although such immobilized enzymes retain their activity, direct electron transfer between these immobilized enzymes and their electrode have not yet been reported.

1.3.2 Electrochemical Regeneration of Coenzymes

Regeneration of cofactors has received considerable attention in connection with the use of immobilized enzymes in bioreactors that involve such cofactors as

NAD(P) and ATP. In the case of enzymatic oxidation coupled with NAD(P), the latter is enzymatically reduced to NAD(P)H. If the enzyme were immobilized on an insoluble matrix, it would be possible to use the immobilized enzyme in a continuous-flow reactor. For further use of the coenzymes, however, NAD(P)H would have to be oxidized to NAD(P). Unless regenerated to NAD(P), the NAD(P)H would have to be taken out of the reaction system, which would have to be recharged with fresh NAD(P).

Several processes for regenerating NAD(P) from NAD(P)H and *vice versa* have been proposed. Both chemical and enzymatic regeneration methods have been applied, which will be described in detail in Chap. 2. As additional chemicals and catalysts (or enzymes) are required, however, these methods are troublesome. In contrast, electrolytic regeneration could offer significant advantages in regeneration of such redox coenzymes as NAD(P) and NAD(P)H, since this works as a continuous-flow process without consuming any other chemicals.

NAD$^+$ and NADP$^+$ have been electrolytically regenerated from the corresponding reduced forms while retaining their coenzymatic function [119, 120]. These electrolytic processes have found to be applicable to cofactors with the following advantages: As electrons transfer directly from a reduced form of the cofactor to an electrode, the electrolytic regeneration requires no additives except a pair of electrodes. In addition, an oxidized form of cofactor can be regenerated in a continuous flow system by applying electric power.

A very special advantage is to electrolytically regenerate the reduced form of nicotinamide cofactors (NADH and NADPH) from the corresponding oxidized form. Reductive electrochemical regeneration of NADH from NAD$^+$, however, although it has been attempted previously by others [121, 122], has not been entirely successful in that electrolytic reduction caused a loss in enzymatic activity. It was generally supposed that dimerization of the NAD radical, produced as an intermediate during electrolysis, prevented the further reduction of NAD$^+$ to NADPH; the dimeric form so produced is enzymatically inactive as shown in Fig. 1.12. The intermolecular coupling of the NAD radical during electrolysis was retarded by electrolytically reducing NAD$^+$ over a mercury electrode coated with a thin liquid crystal membrane of cholesteryl oleate and thereby producing enzymatically active NADH [123]. From a practical viewpoint, however, the liquid-crystal-coated mercury electrode is not suitable for use in a continuous reactor.

Immobilization of NAD$^+$ to polymer matrices could be expected not only to prevent the intermolecular interaction of the NAD$^+$ molecules, which is a prerequisite for dimerization, but also to offer many of the desirable characteristics of the usual immobilized enzyme catalysts. NAD$^+$ has been immobilized covalently onto alginic acid using 1-cyclohexyl-3-(2-morpholineothyl)-carbodiimide metho-*p*-toluenesulfonate. Reductive electrolytic regeneration of NADH from NAD$^+$ immobilized in this manner onto alginic acid was conducted over a mercury pool electrode or a stainless-steel gauze electrode. The bound NAD$^+$ was characterized by cathodic polarography producing a two-step reduction wave in the presence of tetraethyl ammonium salt. The half-wave potential of each reduction wave for the bound NAD$^+$ was slightly shifted to negative values, as compared to those of free NAD$^+$. Controlled-potential electrolysis was conducted for regenerating the reduced form (NAD-alginic acid) of the coenzyme-alginic acid

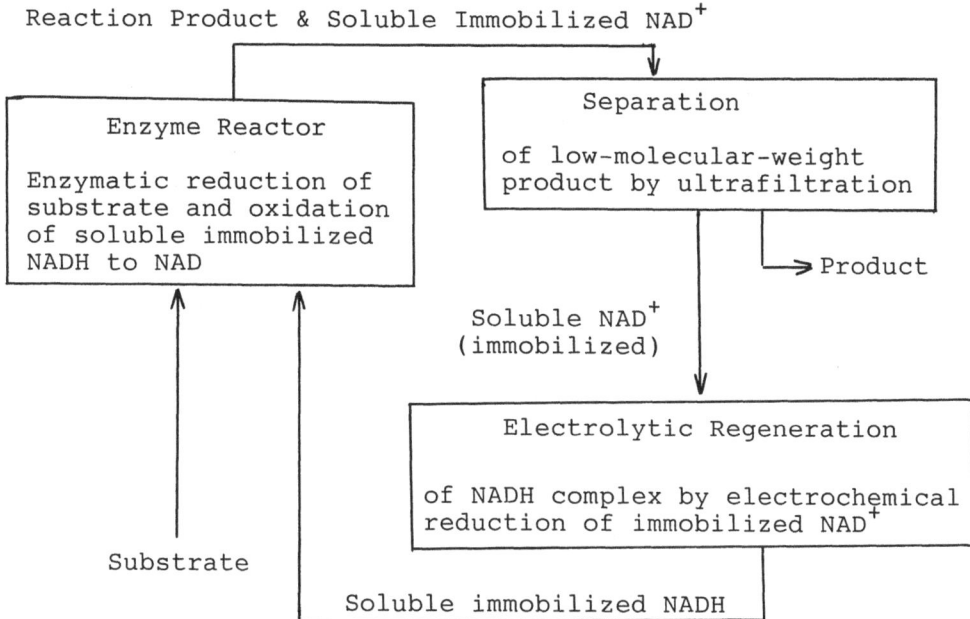

Fig. 1.21. Electrolyte regeneration of NADH from NAD$^+$

complex from the oxidized form (NAD$^+$-alginic acid). The cathodic potential was controlled at a potential of -1.75 V *vs.* SCE, a value slightly more negative than the half wave potential of the second reduction wave. During electrolysis, an increase in absorbance at 340 nm was observed for the catholyte due to production of the reduced form of the NAD complex. The electrolytically regenerated NADH was proved to retain the enzymatic activity.

The electrolytic regeneration of NADH from NAD$^+$ can be employed in continuous-flow enzyme reactors in a manner analogous to similar reaction systems employing electrolytic regeneration of NAD$^+$ from NADH. Continuous analytical schemes using NADH can be based on such reaction systems which regenerate coenzymes and thereby operate continuously without any addition of coenzyme or other chemicals nor change of pH. Figure 1.21 shows a schematic diagram of such a feasible cyclic process consisting of an enzyme reactor utilizing insoluble immobilized apoenzyme and soluble immobilized cofactor with continuous electrolytic regeneration of the reduced form of cofactor and continuous separation of low-molecular-weight reaction products through an ultrafiltration membrane.

1.3.3 Electric Control of Enzyme Activity

Enzymes in metabolic pathways can be maintained at precise levels so that substrates or intermediates can in turn be controlled at physiologically proper concentrations by compartmentalization of enzymes, by controlling the alternate pathways of metabolism, by feedback controls of a number of different modes,

Photocontrol

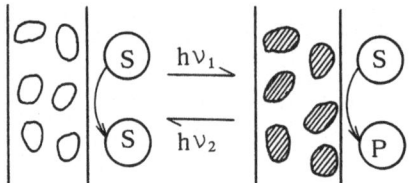

Electric control

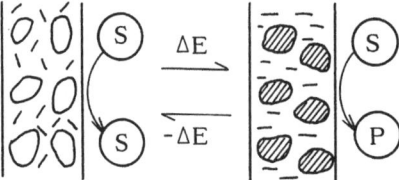

Temperature control

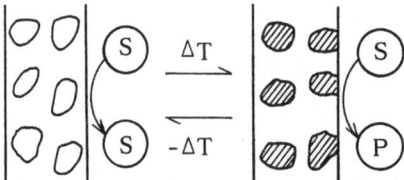

Fig. 1.22. Control of enzyme activity

by chemical modification of enzymes, by involving cascade systems, by proteo-lytic modification of enzymes and by repression and induction. The behavior of many membrane enzymes is markedly affected by their insertion into the membrane structure. Neighboring molecules (e.g., phospholipids and proteins) may play a major role in regulating the catalytic activity of these enzymes. Changes in membrane phospholipids may therefore serve a major regulatory function. A number of regulating enzymes are controlled in that the effectors induce conformational changes in the protecting structures which physically modify the catalytic site of the enzyme. All of these suggestions are profitable in regulating membrane-bound enzyme activity.

These regulatory systems in vivo, however, are too sophisticated to be applied for practical use. Several methods have been proposed to control the enzyme activity, which might be feasible in practical applications. The concepts of these methods are schematically illustrated in Fig. 1.22.

1.3.3.1 Photocontrol of Enzyme Activity

It has been shown by SUZUKI et al. that chemical modification by spiropyran compounds can make enzymes photoresponsive [124, 125]. A number of spiro-pyran compounds undergo photochromism as illustrated in Fig. 1.23; that is, the

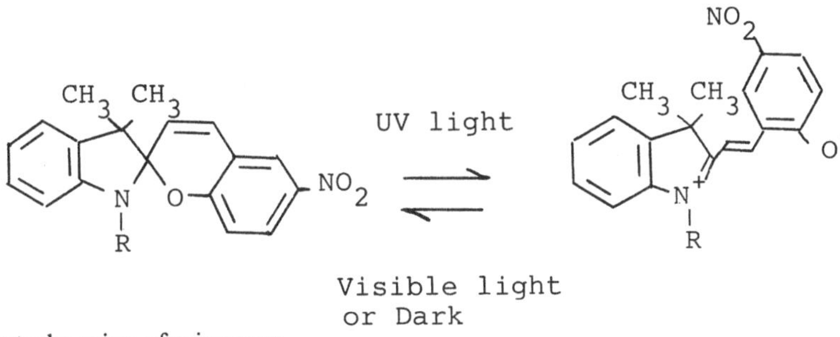

CH$_3$ CH$_3$ UV light   CH$_3$ CH$_3$

N   O—   NO$_2$ ⇌   N$^+$
| |
R   Visible light   R
           or Dark

NO$_2$

O$^-$

Fig. 1.23. Photochromism of spiropyran

reversible photoisomerization of colored (open ring structure) and colorless (closed ring structure) forms. Several enzymes such as amylase were modified with a spiropyran compound to produce photoresponsive enzyme activity. The mechanism of photoregulation was indicated by the drastic change in polarity due to photoisomerization of enzyme-bound spiropyran.

Photocontrol of enzyme activity also has been demonstrated with membrane-bound enzymes. Urease was modified with a spiropyran and immobilized on a collagen membrane matrix. The activity of the modified urease decreased with ultraviolet irradiation and then was restored to the initial activity with visible light [126]. A collagen membrane matrix also has been modified with a spiropyran compound rendering it photosensitive. Trypsin was immobilized on the photosensitive collagen with a slight decrease by about 20% in the dark [127]. The spiropyran-modified membrane was hydrophilic in the dark and turned hydrophobic under visible light, which may have been responsible for the photo change in the apparent diffusion coefficient of the substrate within the membrane matrix.

All of these facts suggest that enzyme activity is controllable by a change of environmental conditions with an external energy source.

1.3.3.2 Electric Field Control of Enzyme Activity

An electric field is expected to be a powerful external source for enzyme activity control. Because liquid crystals are extremely sensitive to an electric field, a liquid crystal supported by a collagen membrane was selected as a membrane matrix for the immobilization of an enzyme to be activity-controlled by an electric field [128]. Lipase was entrapped in a collagen membrane containing 4-methoxyben-zilidene-4'-n'-butylaniline (MBBA). The lipase-liquid crystal membrane was prepared by casting a suspension of collagen fibrils, MBBA, and lipase on a Teflon plate and drying at 20 °C. The membrane was treated with a 0.1% glutaraldehyde solution for 30 s, washed, and dried. An electric field was applied across the lipase-liquid crystal membrane fixed on a platinum electrode. The membrane-bound lipase showed that the activity depended on the electric field.

1.3.3.3 Temperature Control of Enzyme Activity

The liposomal membrane, consisting of phospholipids, causes a drastic change in membrane characteristics at the phase transition temperature. An "enzyme switch," in which the enzyme activity is alternatively changed with an external, thermal source, has been developed by MATSUOKA et al. using a liposome which incorporates a conjugate of a purple membrane and an enzyme [129]. Urease was covalently bound to the purple membrane of *Halobacterium halobium* while retaining its activity. The purple membrane-urease conjugate was incorporated into liposomes of dipalmitoyl phosphatidyl choline by sonication. The liposome-bound urease showed no appreciable activity below a transition temperature of 42 °C. By elevating the temperature above the transition point, the enzyme activity sharply increased indicating that urease activity could be switched on and off repeatedly by temperature changes.

1.3.3.4 Conductive Enzyme Membrane

AIZAWA et al. have recently proposed a novel system to control enzyme activity [93]. The system consists of a conductive polymer membrane in which enzyme molecules are immobilized. Enzyme-immobilized conductive membranes have been prepared by electrochemical polymerization of pyrrole in the presence of enzyme molecules, e.g., glucose oxidase (GOD) and alcohol dehydrogenase (ADH). Membrane-bound GOD, for instance, was found to be electrochemically regenerated from its reduced form, which indicated a possible electron transfer from GOD to the electrode through polypyrrole. Moreover, the enzyme activity of membrane-bound ADH drastically changed depending on the controlled potential.

An electrochemically synthesized polypyrrole membrane has become of interest due to both its high conductivity, which is controlled by electrochemical doping and undoping, and its simple preparation in an aqueous phase. Many applications of electrochemically synthesized polypyrrole membranes have thus been proposed, involving electrochromic displays (ECD) and polymer batteries. Ion gating, ion sieving, and the electrochemical release of neurotransmitters also have been demonstrated with an electrochemically synthesized polypyrrole membrane [130, 131]. Electrochemical polymerization of pyrrole has been applied to synthesize a conductive enzyme membrane which retains its enzyme activity as well as electric conductivity.

Conductive enzyme membranes were synthesized by electrochemical polymerization of pyrrole in the presence of enzyme: a polypyrrole-thin-membrane-deposit on the electrode surface entrapping the enzyme-in-membrane matrix. The membranes were found to remain conductive even when containing enzyme molecules in the membrane matrix.

According to the voltammetry of pyrrole in the presence of GOD, the electrode potential was controlled at 1.0 V *vs.* Ag/AgCl for the electrochemical synthesis of a PP-GOD membrane. The PP-GOD membrane was rinsed with citrate buffer (pH 5.3) after electrochemical preparation, and was prepared with $30 \, mg \cdot ml^{-1}$ of GOD solution. Membrane-bound GOD showed $3 \, units \cdot cm^{-2}$ of enzyme activity. The enzyme activity of membrane-bound GOD sharply depended on the GOD concentration of the electrolyte solution. An increase in enzyme activity with GOD concentration may result from an increase in the total amount of membrane-bound GOD per unit area.

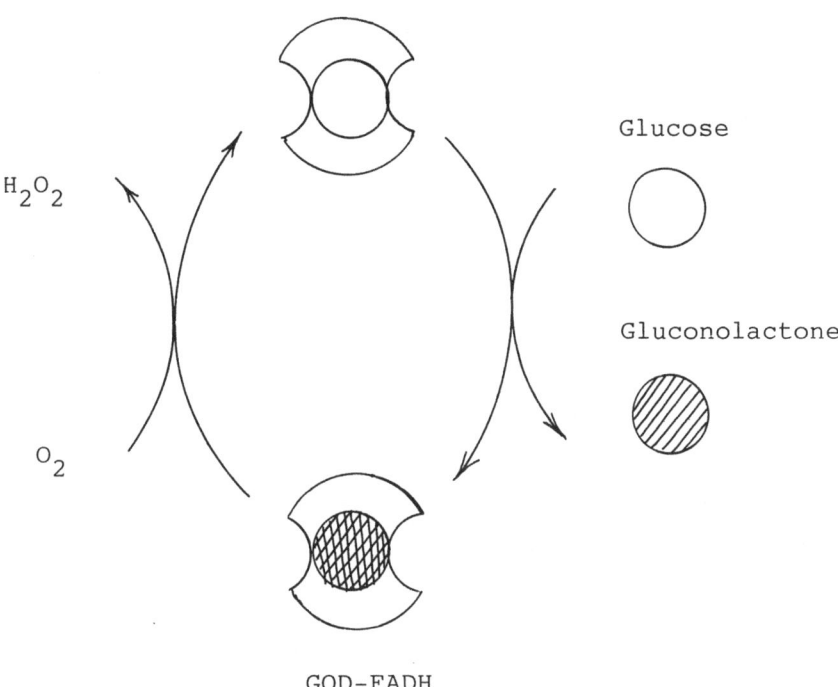

GOD–FAD

Glucose

H$_2$O$_2$

Gluconolactone

O$_2$

GOD–FADH

Fig. 1.24. Reaction scheme of glucose oxidase

The potential-controlled electron transfer was performed with a polypyrrole-glucose oxidase (PP-GOD) membrane. Glucose oxidase (GOD) (E.C. 1.1.3.4) catalyzes oxidation of β-D-glucose in the presence of dissolved oxygen as shown in Fig. 1.24. GOD is reduced when glucose is oxidized. Oxidized GOD may be regenerated from its reduced form by oxygen. The redox reaction of GOD is conducted with a cofactor of flavin adenine dinucleotide (FAD). If the reduced form of GOD need to be electrochemically active in order to be oxidized, glucose should be consecutively oxidized without oxygen by GOD which is coupled with an electrochemical reaction.

It is noted that the reduced form of membrane-bound GOD showed a distinctive anodic peak in the differential pulse voltammogram; membrane-bound FAD is also involved (Fig. 1.25). This result strongly supports the possibility of an electrochemical regeneration of the oxidized form of membrane-bound GOD from its reduced form.

A PP-GOD membrane has been immersed into a solution containing glucose. Dissolved oxygen was carefully evacuated from the solution. The potential of the PP-GOD membrane was controlled with a potentiostat. A steady state of current was measured at each potential. A current-potential curve was also determined in the absence of glucose. The difference between these two curves clearly showed

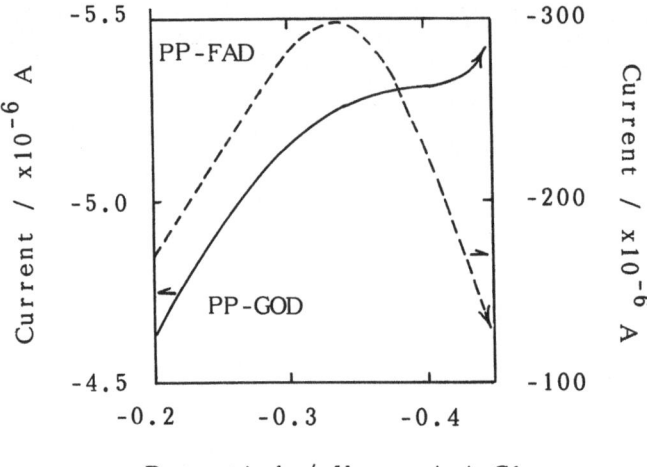

Fig. 1.25. Differential pulse voltammogram of polypyrrole-GOD membrane
Scan rate: 2 mV s^{-1}
Pulse: 50 mV
             50 ms
Electrolyte: 1 *M* KCl
             0.1 *M* Citrate buffer, pH 5.3

electrochemical oxidation of glucose due to the possible scheme of electrochemical regeneration of the oxidized form of membrane-bound GOD from its reduced form. Potential-controlled oxidative electrolysis, which was conducted with a glucose solution and membrane-bound GOD, showed strong support for the above interpretation.

1.3.3.5 Electrochemical Control of Enzyme Activity

Alcohol dehydrogenase (ADH) was also entrapped in a polypyrrole membrane by the electrochemical method. The membrane-bound ADH showed enzyme activity when it was reacted with ethanol and NAD$^+$ in solution. The products may diffuse out from the membrane not to disturb the further reaction. The diffusion of these substrates and products might be influenced by the molecular environment of membrane-bound ADH. The membrane-bound ADH was assayed for its enzyme activity at different electrode potentials. The potential dependence of enzyme activity is shown for membrane-bound ADH in Fig. 1.26. It was shown that enzyme activity at 0.7 V *vs.* Ag/AgCl was more than ten times higher than at 0.1 V *vs.* Ag/AgCl. The enzyme activity was found to markedly depend on the potential of the membrane.

Figure 1.27 represents Lineweaver-Burk plots for membrane-bound ADH at controlled potentials of 0.1 and 0.7 V *vs.* Ag/AgCl. Michaelis constants at 0.1 V and 0.7 V were determined as 5.9×10^{-3} and 8.0×10^{-3} *M*, respectively. The maximum velocity at 0.7 V *vs.* Ag/AgCl was more than ten times higher than at 0.1 V *vs.* Ag/AgCl. Unexpectedly there was no appreciable dependence of the Mi-

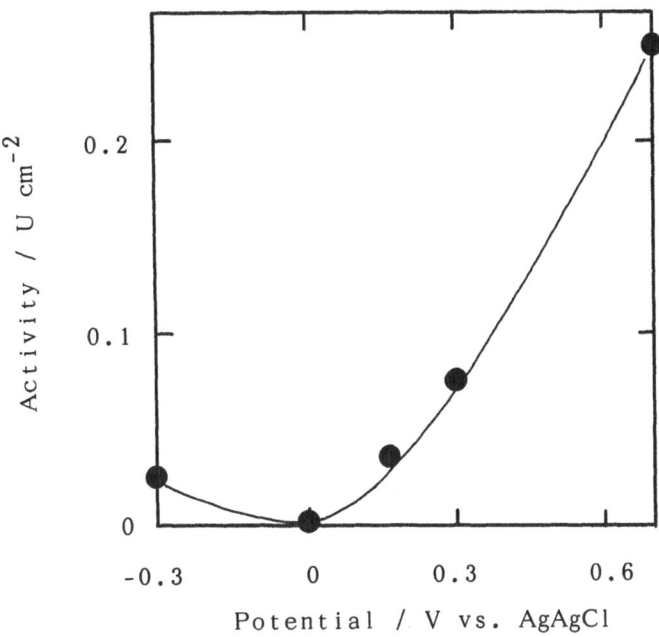

Fig. 1.26. Potential dependence of enzyme activity of polypyrrole-bound ADH
0.66 M Ethanol
4 mM NAD
0.1 M Phosphate buffer, pH 7.5

Fig. 1.27. Lineweaver-Burk plots for polypyrrole-bound ADH at different potentials
○ 0.15 V $vs.$ Ag/AgCl
$K'_m = 5.9 \times 10^{-3}\,M$, $V'_{max} = 7$ mU·cm^{-2}
● 0.70 V $vs.$ Ag/AgCl
$K'_m = 8.0 \times 10^{-3}\,M$, $V'_{max} = 72$ mU·cm^{-2}

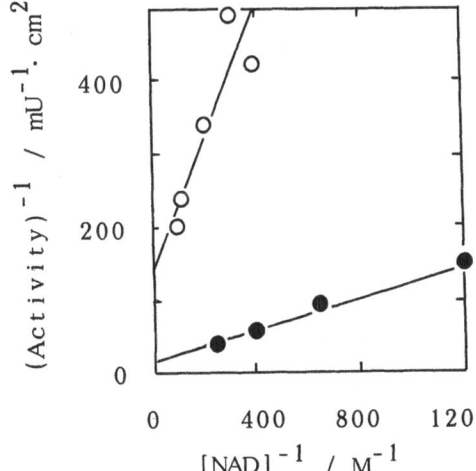

chaelis constant on the controlled potential. Although the substrate affinity of membrane-bound ADH remains constant, the diffusion properties of substrates, especially NAD^+, should be enormously influenced by the controlled potential. The net charge of the membrane matrix drastically changes with the controlled potential.

Potential-controlled enzymatic oxidation of ethanol was pursued with a PP-ADH membrane in the presence of ethanol and NAD^+. NADH increased rapidly at a controlled potential of 0.6 V *vs.* Ag/AgCl. In contrast to this, an increase in absorption at 340 nm was intensively depressed at a controlled potential of 0 V *vs.* Ag/AgCl.

It has clearly been demonstrated that the enzymatic reaction is regulated by the potential of conductive enzyme membranes.

1.4 Electroanalytical Application of Enzymes

1.4.1 Electrochemical Enzymatic Assay

An enzyme is capable of catalyzing a particular reaction of a particular substrate to produce a characteristic and measurable reaction end product. This specificity of enzymes and their ability to catalyze reactions of substrates at extremely low concentrations is of significant importance in chemical analysis. Enzymes are also highly selective with respect to particular reactions.

Enzyme-catalyzed reactions have long been used for analytical purposes in the determination of substrates, activators, inhibitors, and also of the enzymes themselves. Because of the relatively recent availability of larger numbers of highly active enzyme preparations at reasonable costs, it is now possible and practical to use enzymes as analytic reagents. Moreover, the methods developed for the immobilization of enzymes onto solid matrices considerably increases the versatility of their use and thus also their economic availability.

Electrochemical methods are also often used for analytical purposes. The development and application of ion-selective electrodes (ISEs) continues to expand in various areas of analytical research. The membrane-covered oxygen electrode developed by CLARK has become one of the more important analytical tools. Current electroanalytical chemistry has become more refined and has developed more powerful and more reliable instrumentation.

The fundamental principle of electrochemical enzyme analysis consists in the utilization of the specificity of enzymes in conjunction with electroanalytical instrumentation.

1.4.1.1 Potentiometric Determination of Enzyme Activity

In the situation where the chemical reaction involves an electroactive reactant, reagent, or product, and where there is an electrode available which can come into Nernstian or quasi-Nernstian equilibrium with the electroactive species, potentiometric measurement can be used. As a result of the extensive development in recent years of solid and liquid membrane electrodes which respond to ionic species and diffusible gases, the variety of electroactive species which can thus be measured analytically has greatly increased.

The most common potentiometric method used in enzymology is the glass electrode in following the production of acids. Because changes in the pH affect the activity of the enzyme and also the rate of reaction, direct readings of pH changes are generally not used. Instead a "pH stat" method is employed, in which the pH is maintained at a constant value by frequent addition of alkali. The rate at which base is added then gives the reaction velocity independent of the amount of buffer.

Potentiometric gas-sensing membrane electrodes and other potentiometric electrodes have been coupled with biocatalyzed reactions to determine either the enzyme, substrate, cofactor, prosthetic group, inhibitor, or activator.

Urea is specifically determined using urease:

$$CO(NH_2)_2 + H_2O \xrightarrow{\text{Urease}} CO_2 + 2NH_3 . \tag{1.18}$$

Enzymatically generated ammonia is measured to determine urea by various methods. Of these methods, the ammonia gas electrode method has been evaluated. The ammonia liberated in the enzymatic reaction can be monitored continuously, and the rate of its production is proportional to the urease concentration. The advantage of this method is that the air gap electrode and the polymer-membrane electrode-based gas electrode suffer no interference from Na^+ and K^+ present in the sample.

The ammonia liberated in the oxidation of amino acids by D-amino acid oxidase is also monitored with an ammonia gas electrode. Plots of E/t are proportional to the concentration of enzyme over the range 10^{-1} to 10^{-3} units.

Several applications of carbon dioxide gas electrodes have been recently reported, which includes the enzymatic determination of thiamine pyrophosphate [132] and the kinetic determination of the proteolytic enzymes leucine aminopeptidase [133] and trypsin [134]. Moreover, the carbon dioxide gas electrode can be coupled with malate dehydrogenase to determine $NADP^+$ and glutathione reductase [135]. Also, the carbon dioxide electrode has been employed to determine vitamin B_6 (pyridoxal phosphate) using tyrosine decarboxylase [136].

Exploration of the use of electrochemical detectors in continuous flow systems, "Flow Injection Analysis" (FIA), is one of the most active areas of analytical chemistry. MASCINI et al. have reported the use of immobilized creatininase in conjunction with a flow-through ammonia electrode for the automated determination of creatinine in blood and urine [137]. The immobilized enzyme converts creatinine in the sample to ammonia and N-methylhydantoin. Background endogenous levels of ammonia in the samples must first be measured without the enzyme and these values are then substracted from the total ammonia detected after the enzymatic process in order to accurately determine creatinine levels.

MOTTOLA et al. have proposed the flow injection determination of penicillins using immobilized penicillinase in a single bead string reaction in conjunction with a pH glass electrode (Fig. 1.28) [138]. Enzymatic determinations of penicillins are based on monitoring directly or indirectly the amount of penicilloic acid resulting from the enzymatic hydrolysis of penicillins:

$$\text{Penicillin} + H_2O \xrightarrow{\text{Penicillinase}} \text{Penicilloic acid} . \tag{1.19}$$

Immobilization on a glass matrix, after ammonium bifluoride etching, using 95% ethanol, aminophenyl trimethoxy silane, and glutaraldehyde coupling provides

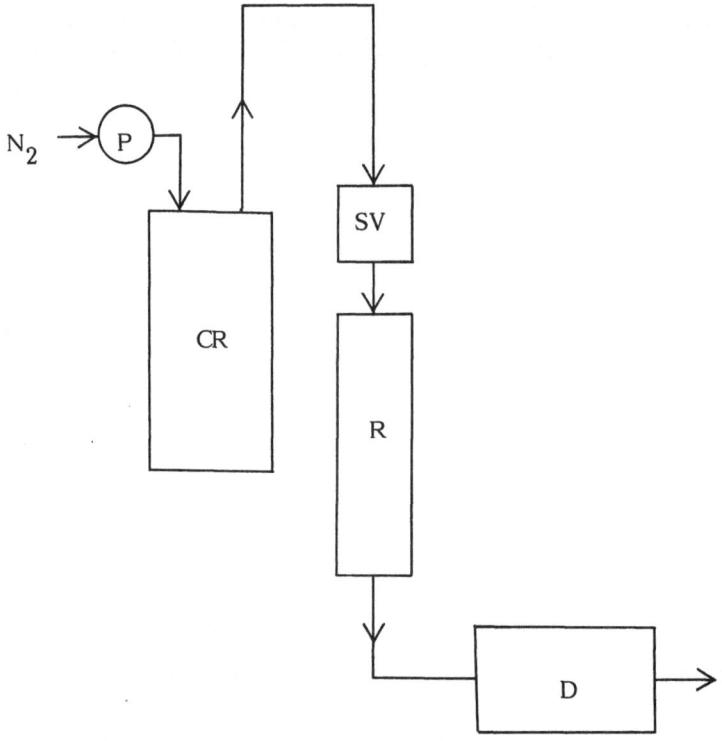

Fig. 1.28. Flow injection analysis system coupled with an enzyme reactor. P: pressure regulator, CR: carrier reservoir, SV: sample injection valve, R: immobilized enzyme reactor, D: detector

preparations with relatively high activity, retaining 97% of the initial activity after 10 months of daily usage. Packed column reactors generally produce undesirable back-pressure problems in continuous-flow sample processing. Single bead string reactors with enzyme immobilized on the single beads provide an excellent configuration for using immobilized enzymes in flow injection systems [139]. Local enzyme activity can be maintained at satisfactory levels and flow characteristics are advantageous. Penicillins G and Y can be selectively determined in the 0.05–0.50 mM range and at a frequency of 150 injections. These analysis techniques become more and more popular with growing automation; the use of ion selective electrodes (ISE) as detectors in such systems also expanding. FIA methods have been applied to clinical chemistry often with the use of ISE-based systems [140].

The activity of several enzymes can be assayed by potentiometric methods, which include the followings (Table 1.9).

Cholinesterases. An electrochemical method for the determination of cholinesterase, based on the hydrolysis of acetylthiocholine iodide by cholinesterase has been proposed. A small, constant current of 25 μA is applied across two platinum electrodes, and the change in potential with time upon hydrolysis is recorded. Ini-

Table 1.9. Potentiometric determination of enzyme activity

Electrode	Enzyme
NH$_3$ electrode	D-Amino acid oxidase
CO$_2$ electrode	Leucine aminopeptidase Trypsin Gluthathione reductase
Fluoride electrode	α-Chymotrypsin
Cyanide electrode	β-Glucosidase
Iodide electrode	Glucose oxidase

tially, a constant potential is obtained due to the formation of the more electrochemically active thiol.

$$(CH_3)_3N-(CH_2)_2-S-\underset{O}{\overset{\|}{C}}-CH_3 \xrightarrow{ChE} (CH_3)_3N-(CH_2)_2-SH. \qquad (1.20)$$
$$\underset{E_0 \sim 0.45\,V}{I^-} \qquad\qquad\qquad \underset{E_0 \sim 0.2\,V}{I^-}$$

The slope of the resulting depolarization curves provides data on the rates of enzymatic hydrolysis of the thiocholine esters. These rates correspond well with those predicted by Michaelis-Menten kinetics. By this procedure 0.2 to 14 U of cholinesterase could be assayed with a standard deviation of 0.7%.

α-Chymotrypsin. A very interesting electrochemical method for monitoring the activity of chymotrypsin is based on the stoichiometric 1:1 molar reaction of the enzyme with diphenylcarbonyl fluoride. An ion-selective electrode is then used to measure the fluoride and hence the chymotrypsin.

Glucose Oxidase. An automated electrochemical method has been developed for the determination of glucose oxidase based on the oxidation of glucose to peroxide, followed by the oxidation of iodide to iodine in the presence of molybdate as a catalyst:

$$\text{Glucose} + O_2 + H_2O \xrightarrow{\text{glucose oxidase}} \text{Gluconic acid} + H_2O_2, \qquad (1.21)$$

$$2I^- + H_2O_2 + 2H^+ \xrightarrow{\text{ammonium molybdate}} I_2 + 2H_2O. \qquad (1.22)$$

The rate of production of iodine is proportional to the rate of oxidation of glucose and is detected either potentiometrically or amperometrically. In either case, automatic control equipment provides a direct read-out of the time required for a predetermined amount of iodine to be produced. The reciprocal of the time interval is proportional to the glucose oxidase activity, with a relative standard deviation of about 2%. Ferrocyanide may be used in place of iodide.

β-Glucosidase. The method is based on the liberation of cyanide from the substrate amygdalin. The cyanide produced is measured using a cyanide electrode.

1.4.1.2 Amperometric Enzyme Assay

Oxygen, hydrogen peroxide, and NAD(P)H are electrochemically active species which are most commonly associated with enzyme reactions. The Clark-type oxygen electrode and hydrogen peroxide electrode thus provide effective tools for amperometric enzyme assays. Amperometric determinations of enzyme activity are listed in Table 1.10.

The oxygen electrode consists of a gold or platinum cathode separated from a silver anode which is housed in a plastic or glass casing and comes in contact with a test solution only through an oxygen permeable polymer membrane (Fig. 1.29). When oxygen diffuses through the membrane it is electrochemically reduced at the cathode by an applied potential of -0.6 V. This reaction causes a current to flow between the anode and cathode which is proportional to the partial pressure of oxygen in the solution.

The oxygen electrode has been increasingly used in the enzymatic analysis of oxygen-consuming enzymatic systems. The determination of glucose upon cata-

Table 1.10. Amperometric determination of enzyme activity

Electrode	Glucose oxidase
NADH electrode	Lactate dehydrogenase Alcohol dehydrogenase Malate dehydrogenase
H_2O_2 electrode	Glucose oxidase

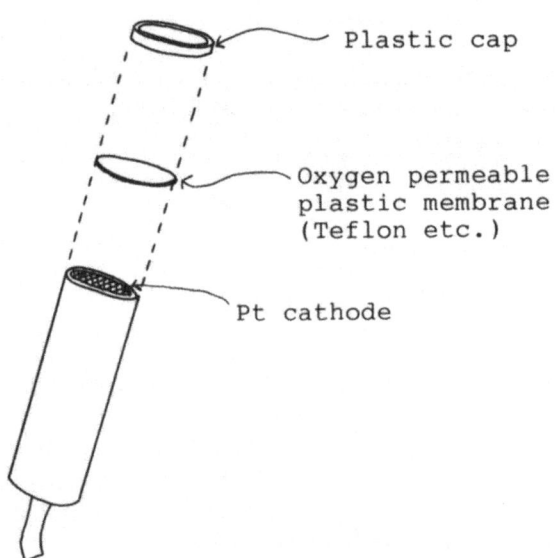

Plastic cap

Oxygen permeable plastic membrane (Teflon etc.)

Pt cathode

Fig. 1.29. Clark-type oxygen electrode

Table 1.11. Amperometric assays of NADH

Electrode	Amperometry	Detection limit of NADH
NAD-adsorbed carbon	Cyclic voltammetry	$10\,\mu M$
Glassy carbon	Flow amperometry	$0.1\,\mu M$
Mercury	Square-wave voltammetry (reduction of NADH)	$7\,nM$

lysis by glucose oxidase continues to provide the most important "test system" for amperometric methods.

Direct electrochemical methods for NADH determination have been developed based on its oxidation at solid electrodes [39, 42, 43, 53, 49, 141–145]. Reduced nicotinamide adenine dinucleotide is detectable with cyclic voltammetry upon oxidation at a carbon electrode coated with adsorbed NAD^+ [146]. These amperometric methods apparently have a detection limit of $10\,\mu M$ [42, 145]. Interferences may limit the effectiveness of these methods in the application to the samples of biological origin [147, 43].

The flow amperometric detection of NADH at a glassy carbon electrode has been shown to be feasible. Reduced nicotinamide adenine dinucleotide is oxidized to NAD^+ by applying a sufficiently positive potential to an electrode. The detection limit for this system is $0.1\,\mu M$.

SHAH and OSTERYOUNG have recently shown the usefulness of rapid scan square-wave voltammetry for the direct determination of NADH at low concentrations (Table 1.11) [148]. This method differs in its concept from other electrochemical methods in that NADH is being reduced at a mercury electrode. This method assures a detection limit of below $7\,nM$. The electrode process involves the totally irreversible reduction of acid-decomposed NAD^+. Square-wave voltammetry has the advantage of small background currents, good sensitivity, low detection limits, and short measurement times. Cathodic stripping voltammetry allows the electrochemical determination of NADH at low concentrations since the adenine-containing compounds are adsorbed easily on the electrode surface [32].

The above-mentioned method assures reproducible sensitivity in the determination of NADH after separation of the reduced coenzyme. For example, blood alcohol has shown to be detectable with amperometry based on liquid chromatography and anodic detection of NADH [149].

Electrochemical techniques coupled with enzymatic reactions expand the applicability of NAD(P)H amperometry; enzymatic activity is determined by electrochemical detection of the reduced coenzyme [51, 150–152].

WALLACE et al. [43] have reported a highly sensitive electrochemical assay for lactate dehydrogenase, alcohol dehydrogenase, and malate dehydrogenase by amperometry at a stationary platinum electrode in a stirred electrolyte. The rate of increase in net anode current in substrate solutions containing as little as 2×10^{-3} units of enzyme per milliliter correlated well with the rate of change in absorbance at 340 nm for each sample [43]. A carbon paste electrode modified chemically with NAD^+ has been used for the determination of ethanol and lactate [153].

1.4.2 Electrochemical Immunoassay

The interaction between antigen and antibody molecules may be extremely specific under favorable conditions as a consequence of binding geometries. Immunoassays, which rely on high selectivity and sensitivity by the molecular recognition of antibodies, are classified into two categories: labeling and non-labeling. Since the pioneering work of YALOW and BERSON, radioisotopes have been employed as the label of choice. However, a number of nonisotopic alternatives have been developed including electron spin resonance for detecting radical labels, nephelometry, fluorescence, chemiluminescence, and enzyme labels, among others. Moreover, immunoassay with an electroactive label has become attractive due to the wide dynamic range and low detection limits of modern electroanalytical methods.

WEBER et al. utilized ferrocene as an electroactive label in the immunoassay for codeine and morphine [154]. A homogeneous voltammetric immunoassay has been explored in conjunction with continuous flow amperometric detection. Morphine acts as an electroactive antigen and codeine as an electroinactive antigen competing for antibody binding sites. Competition between the two molecules resulted in the release of morphine from the antibody complex with a subsequent increase in the current signal due to unbound morphine. HEINEMAN et al. used a similar approach to determine estriol with the use of Hg-labeled estriol [156]. Estriol and Hg-labeled estriol compete for the limited binding site on the estriol antibody.

GLERIA et al. have recently reported an homogeneous ferrocene-mediated amperometric immunoassay [155]. They used ferrocene and its derivatives to act as electron acceptors for glucose oxidase (GOD) in a non-oxygen-dependent manner as outlined in reactions 1–3, below. The overall process involved

$$GOD_{ox} + glucose \rightarrow GOD_{red} + gluconolactone, \tag{1.23}$$

$$GOD_{red} + 2FeCp_2R^+ \rightarrow GOD_{ox} + 2FeCp_2R + 2H^+, \tag{1.24}$$

$$2FeCp_2R \rightarrow 2FeCp_2R^+ + 2e^-, \tag{1.25}$$

in the homogeneous assay concept is illustrated in Fig. 1.30. The binding of the ferrocene-drug (antigen) complex by antibody inhibits its ability to act as a mediator in the GOD-catalyzed reaction, and thus the catalytic current is greatly decreased. This can be reversed by adding a nonlabeled drug (i.e., analyte) that competes for the available antibody binding sites. Thus, the catalytic current produced in the reaction depends on the concentration of analyte in solution. Lidocaine ((α-diethylamino)-2,6-dimethylacetanilide) (Fig. 1.31) in plasma was determined over the concentration range 5–50 μM with a relative standard deviation of 3–6%.

WEHMEYER et al. have reported the development of a homogeneous voltammetric immunoassay for estriol labeled in the 2- and 4-positions with nitro groups us-

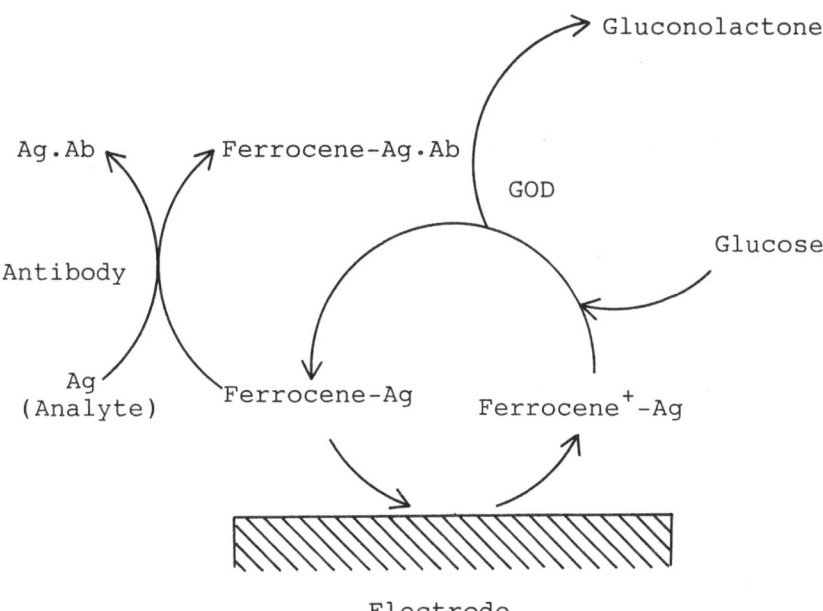

Fig. 1.30. Overall process involved in the homogeneous assay concept

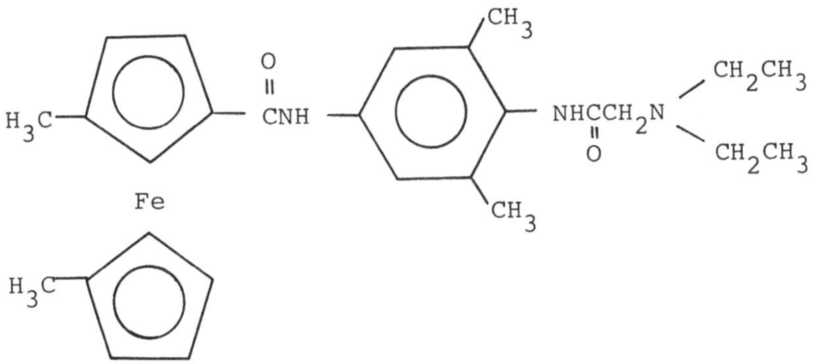

Fig. 1.31. Ferrocene-labeled Lidocaine

ing differential pulse polarographic detection [157]. Immunocomplex formation of 2,4-dinitroestriol with the corresponding antibody was found to result in a decrease in the reduction current.

ALAM et al. have reported a voltammetric immunoassay for human serum albumin in which lead, bound nonspecifically to albumin, serves as an indicator [158]. A heterogeneous voltammetric immunoassay has been developed by HEINE-MAN et al. on the basis of a metal-chelate label using indium as the indicator metal [159]. Human serum albumin (HSA) was chosen as a model antigen. Diethylene triamine pentaacetic acid (DTPA) was coupled to the amino residues of HSA and

49

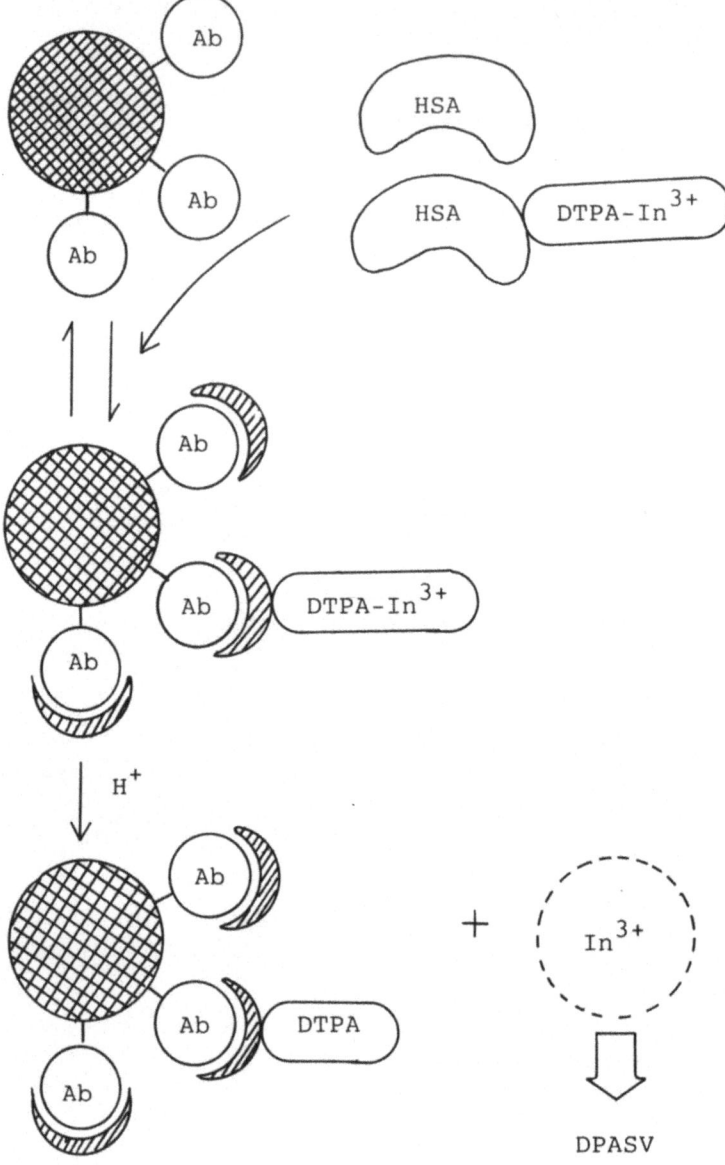

Fig. 1.32. Heterogeneous voltammetric immunoassay on the basis of a metal-chelate label using indium

acted as a site specific chelate agent for metal ions. Indium (In^{3+}) is very tightly bound by DTPA ($K_f = 10^{29}$) making the labeled protein complex stable until the In^{3+} is purposefully released by acidification. The heterogeneous voltammetric immunoassay is based on the competition between labeled and native antigen (HSA) for a limited number of antibody binding sites which have been insolubil-

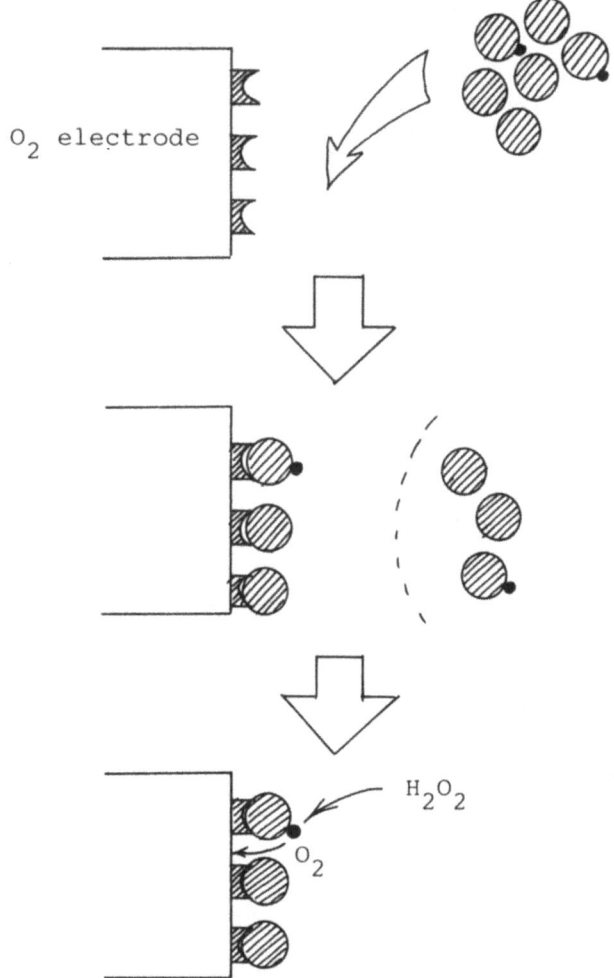

Fig. 1.33. Heterogeneous amperometric enzyme immunoassay using an oxygen electrode

ized on an immunoadsorbent. Following equilibration, the antibody "bound" complexes were released by acidification which were detected separately by differential pulse anodic striping voltammetry (DPASV) (Fig. 1.32). At the molar labeling ratio used, a detection limit for HSA of approximately $5.0 \ \mu g \, ml^{-1}$ was obtained.

The sensitivity and detection limits of electrochemical approaches are enhanced when coupled to chemical amplification mechanisms such as enzyme catalysis. A number of enzyme labels have been monitored amperometrically [160–162].

An amperometric technique has been used by AIZAWA et al. who determined immunoglobulin G (IgG) by a competitive binding technique involving catalase-labeled IgG and IgG [163]. Antibody is covalently bound to a polymer mem-

Table 1.12. Heterogeneous amperometric enzyme immunoassay

Determinant	Receptor	Transducer	Range $(\mathrm{mg \cdot ml^{-1}})$	Remarks
IgG	Anti-IgG Ab membrane Catalase-labeled IgG	O_2 electrode	10^{-4}–10^{-3}	Competitive
	Anti-IgG membrane GOD-labeled IgG	O_2 electrode O_2 electrode	10^{-6}–10^{-3} 10^{-6}–10^{-3}	Sandwich Sandwich
IgM	Anti-IgM Ab membrane Catalase-labeled Ab	O_2 electrode	10^{-7}–10^{-4}	Sandwich
Albumin	Anti-Albumin Ab membrane Catalase-labeled Ab	O_2 electrode	10^{-6}–10^{-3}	Sandwich
HCG	Anti-HCG Ab membrane Catalase-labeled HCG	O_2 electrode	10^{-2}–10^2 IU. ml^{-1}	Competitive
AFP	Anti-AFP Ab membrane Catalase-labeled AFP	O_2 electrode	10^{-11}–10^{-8}	Competitive
HBs	Anti-HBs Ab membrane Peroxidase-labeled HBs	I^- electrode	10^{-7}–10^{-5}	Competitive
Ab	Antigen-labeled liposome	TPA^+ electrode	–	
OTA	Anti-OTA Ab membrane	O_2 electrode	10^{-10}–10^{-5}	Competitive

brane. Catalase-labeled IgG and IgG (analyte) competitively react with membrane-bound antibody. The amount of catalase-labeled IgG fixed to the membrane is determined by measuring amperometrically the oxygen generated by the catalase. A Clark type oxygen electrode was used for amperometric measurement of oxygen. The assay process is presented in Fig. 1.33. AIZAWA et al. have applied the amperometric enzyme immunoassay to determine human chorionic gonadotropin (HCG), α-fetoprotein (AFP), β_2-microglobulin, and ochratoxin A. The characteristics of this amperometric enzyme immunoassay are summarized in Table 1.12 [164]. YUAN et al. also reported an amperometric technique with a platinum electrode for immunoelectrochemical determination of creatine kinase isoenzyme MB [165].

Applications of ion-selective electrodes have been extended to include immunoassays. ALEXANDER et al. reported an enzyme-linked immunoassay method for IgG based on the precipitin technique using a fluoride-selective electrode and horseradish peroxidase (HRP)-labeled anti-IgG [166]. The complete shape of the precipitin reaction curve is determined by specifically precipitating various amounts of human IgG with anti-human IgG labeled with HRP, and the enzyme activity of the washed precipitates is determined by analysis in a continuous-flow electrode system after dissolution of the precipitin in dilute acetate buffer. The analysis is based on the continuous-flow detection of the liberated fluoride after the enzymatic reaction between the HRP–H_2O_2 complex and p-fluoroaniline, resulting in the cleavage of a C–F bond. The concentration of the liberated fluoride is shown to be proportional to the HRP activity.

An ammonia gas electrode was used for determining the urease label in the enzyme immunoassay of cyclic adenosine monophosphate [167].

1.4.3 Enzyme Sensors

Advances in sensor technology have opened the way for new and sophisticated measurement techniques and instrumentation for biotechnological process control, environment control, and medical diagnosis. Among various sensors, a biosensor can offer high selectivity for a specific substance, which has not been attained by synthetic sensor materials. The new breed of biosensors will present a unique combination of matrix-bound biochemical, transducer, and microelectronic components to allow almost instantaneous determination of substrate, analyte, or ligand concentrations.

The basic principles of biosensors are schematically illustrated in Fig. 1.34. Molecular recognition of substrates by enzymes, organelles, microorganisms or tissue slices is followed by conversion into the corresponding products which are detected and recorded by the electronic device. The physico-chemical changes associated with molecular recognition caused by binding of the substance to be determined or by the enzymatic substance conversion are transduced by potentiometric or amperometric electrodes, thermistor, field effect transistors (FET), optoelectronic detectors, optical fibers, or other devices into an electric signal. Biosensors are classified into enzyme sensors, microbial sensors, immunosensors, and others.

Ever since CLARK and LYONS presented the first enzyme electrode in 1962, many workers have been occupied with the development of enzyme electrodes in

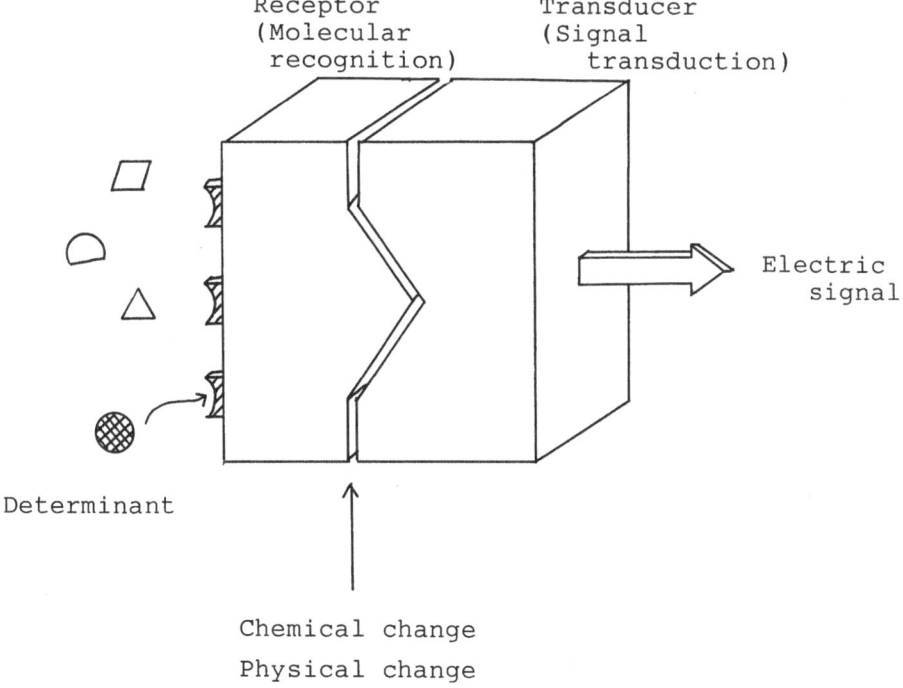

Fig. 1.34. Conceptual illustration of biosensor

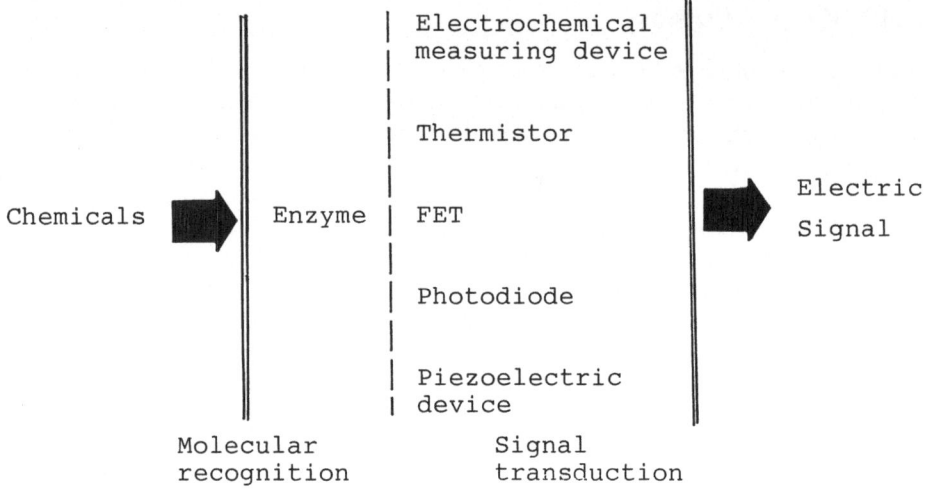

Fig. 1.35. Molecular recognition and signal transduction in an enzyme sensor

general [168–173]. The enzyme electrode, which consists of matrix-bound enzyme and an electrochemical measuring device, has found wide application in various fields. In a similar manner to the enzyme electrode, matrix-bound enzyme recently has been coupled with electronic devices such as a thermistor, field effect transistor (FET), and photodiode to construct biosensors. The term "enzyme sensor" is used to describe these sensors.

The molecular recognition part of enzyme sensor consists of the matrix-bound enzyme which converts a specific substance (determinant) selectively into products. The enzymatic reaction is accomplished by a change in chemicals, photons, and temperature. Such a change is transduced into electric signals at the transducer (Fig. 1.35).

Enzymes are commonly used in a matrix-bound form. Several methods have been established to immobilize enzymes onto the matrix; (1) Covalent, (2) Entrapment, (3) Adsorption, and (4) Encapsulation methods. Polymer membrane matrices are widely used to couple the immobilized enzyme with an electrochemical measuring device. Enzymes should retain sufficient activity in the matrix-bound form. It is required that the polymer membrane matrix should provide permeability to the corresponding substrate and products. The performance of an enzyme sensor may reflect these properties of matrix-bound enzymes.

Enzyme sensors may be classified as:

(1) Enzyme-electrode-type sensors
(2) Enzyme thermistor
(3) Enzyme transistor
(4) Enzyme field effect transistor (ENFET)
(5) Enzyme photodiode
(6) Optical enzyme sensor

The selectivity of all of these enzyme sensors depends on the molecular recognition of the matrix-bound enzyme.

A. Enzyme Electrode

Enzyme Electrodes for Glucose. A glucose sensor installed in a glucose analyzer became commercially available in the mid-1970s. In the early 1980s, extensive development of glucose analyzers and sensors has taken place. These analyzers, which have been developed for clinical use, require 5–20 μl of serum sample for blood glucose analysis.

The basic structure of a glucose sensor is presented in Fig. 1.36. The sensor consists of a glucose oxidase (GOD) membrane and a Clark-type oxygen electrode. GOD selectively catalyzes the following reaction:

$$\beta\text{-D-Glucose} + O_2 + H_2O \ \rightarrow \ \beta\text{-D-Gluconate} + H_2O_2 \,. \tag{1.26}$$

Glucose in a sample solution is oxidized with the resulting consumption of oxygen when contacted with the membrane-bound GOD. The decrease of dissolved oxygen is sensitively detected with the oxygen electrode. The output change of the sensor reflects the concentration of glucose in solution. Typical response characteristics of the enzyme sensor for glucose are presented in Fig. 1.37 [174].

Hydrogen peroxide is formed in the GOD-catalyzed reaction. A GOD membrane may be coupled with a hydrogen peroxide electrode to form another type of glucose sensor. Both oxygen electrode and hydrogen peroxide electrode-based glucose sensors are now commercially available.

Glucose oxidase has been immobilized onto various matrices as shown in Table 1.13 [175]. Gel entrapment and cross-linking with bifunctional agents, e.g., glutar-

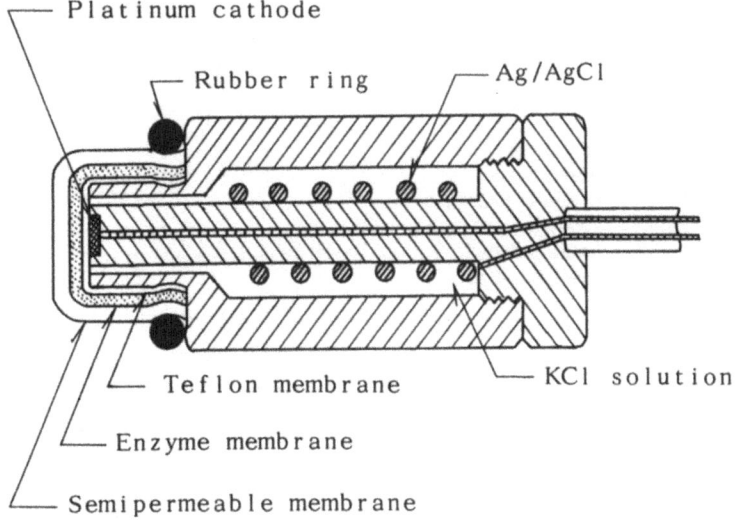

Fig. 1.36. Configuration of an enzyme sensor for glucose

55

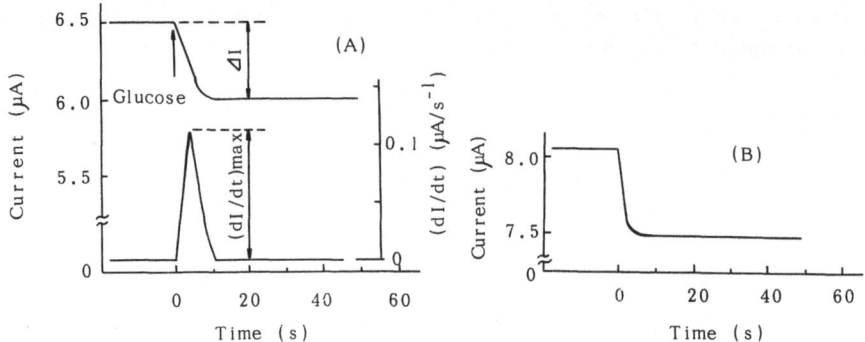

Fig. 1.37. Response characteristics of an enzyme sensor for glucose

Table 1.13. Immobilized enzymes for enzyme sensors

Membrane matrix	Enzyme	Immobil- ization	Trans- ducer	Deter- minant	Response time	Range (mg/L)
Polyacryl amide	GOD	Entrap	O_2 EL.	Glucose	0.5–3 min	10^2–10^3
	ADH	Entrap	Pt EL.	Ethanol	–	1–10^4
	Tyrosinase	Entrap	Pt EL.	Phenol	5–10 min	0.05–10
Gelatin GA*	GOD	Entrap	O_2 EL.	Glucose	15 s	1–5×10^4
Albumin GA	ADH	Crosslink	O_2 EL.	Ethanol	30 s	5–10^3
Collagen	Invertase, GOD Mutarotase	Entrap	O_2 EL.	Sucrose	5 min	10^2–5×10^3
	Catalase	Entrap	O_2 EL.	H_2O_2	2 min	1–100
Polyamide GA	ADH	Covalent	C EL.	Ethanol		
AN*² GA	LOD	Covalent	O_2 EL.	Lactate	30 s	5–2×10^3
CTA*³ Triamine*⁴ GA	GOD	Covalent	O_2 EL.	Glucose	10 s	

*1 GA: Glutaraldehyde.
*2 AN: Acrylonitrile.
*3 CTA: Cellulose triacetate.
*4 Triamine: 1,8-diamino-4-aminomethyl octane.
ADH: Alcohol dehydrogenase, GOD: glucose oxidase.
LOD: Lactate oxidase.

aldehyde, in mixtures with an inert protein are frequently used immobilization methods for enzymes of low specific activity. Polyacrylamide, gelatin, and collagen are representative matrices. The method of covalent coupling, e.g., to nylon netting or collagen membrane, has proven effective for receptors binding high-molecular-weight or surface-active substances which are not able to diffuse into the gel layers. The operation of the electrode requires the diffusion of the analyte from the gas permeable membrane and the immobilization medium to the en-

56

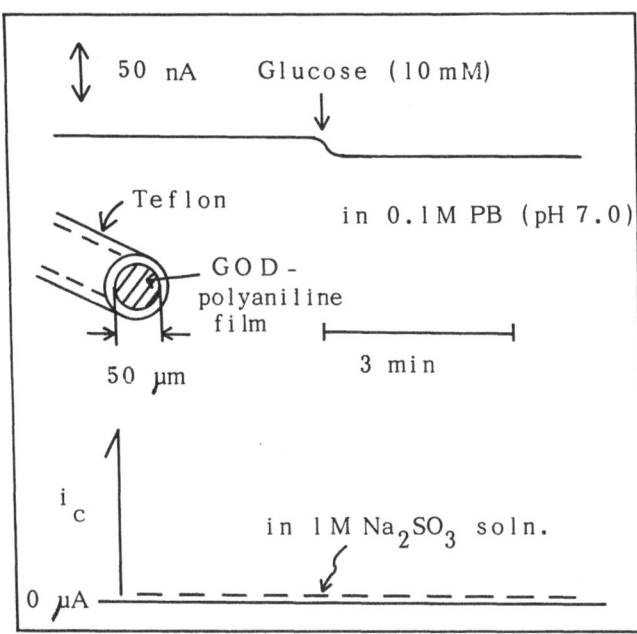

Fig. 1.38. Fabrication of a micro-enzyme sensor for glucose by electrochemical polymerization of aniline and glucose oxidase

zyme, and subsequent detection of the reaction product by the electrode before the product diffuses back into the solution.

SHINOHARA et al. modified the platinum electrode surface with a GOD-entrapped polyaniline thin layer [176]. The modification was accomplished by the electropolymerization of aniline in the presence of GOD in a neutral aqueous solution. The GOD-entrapped polyaniline membrane rejects the permeation of chemicals except, gas molecules such as dissolved oxygen. A microenzyme sensor can be fabricated by the electrochemical polymerization (Fig. 1.38).

Despite the emphasis on amperometric determination of either oxygen or hydrogen peroxide, there are some disadvantages in using an oxygen-coupled glucose assay. Variations in oxygen tension of the sample may introduce fluctuations into the electrode response; at low oxygen tension, the upper limit of linearity for the current response may be reduced. In commercial analyzers, this is usually circumvented by predilution of the sample into oxygenated buffer. HIGGINS et al. developed an alternative amperometric detection method, based on glucose oxidase, that is not dependent on oxygen as a mediator of electron transfer [177]. The electrode uses a substituted ferricinium ion as a mediator of electron transfer between immobilized glucose oxidase and a graphite electrode (Fig. 1.39).

In the condition of diabetes mellitus, the determination of blood glucose levels rapidly, conveniently, precisely, and economically is important for its diagnosis and effective management.

The bedside-type artificial endocrine pancreas, the closed-loop control system used in the treatment of diabetic patients, has revealed that insulin delivery in re-

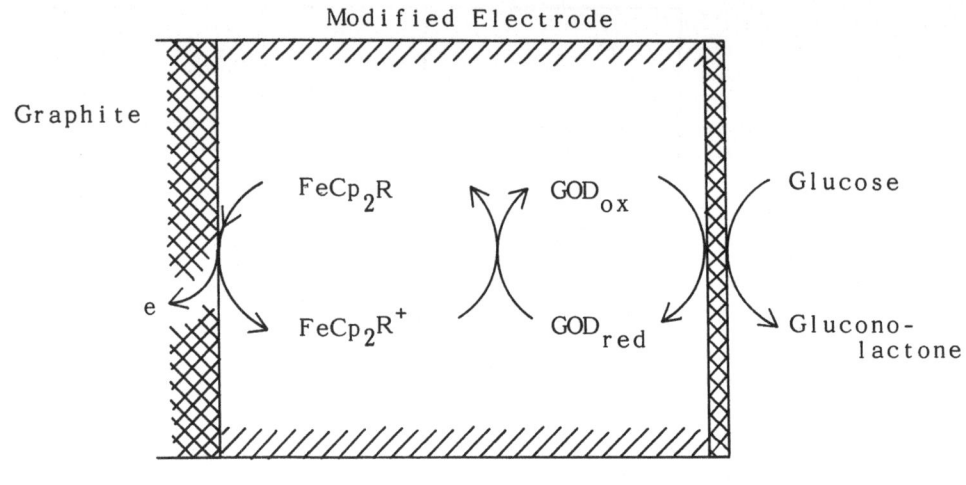

Modified Electrode

Graphite

FeCp$_2$R   GOD$_{ox}$ Glucose

e

FeCp$_2$R$^+$ GOD$_{red}$ Gluconolactone

FeCp$_2$R : ferrocene

GOD   : glucose oxidase

Fig. 1.39. Amperometric enzyme sensor for glucose using an electron mediator

sponse to measured blood glucose concentrations on a minute-by-minute basis can normalize glycemia in diabetic patients following various stimuli. However, the problems associated with long-term venous access and the size of the artificial endocrine pancreas have demonstrated that this system has a limited applicability in clinical practice. The minituarization of a glucose monitoring system is an essential requirement for its clinical application. SHICHIRI et al. have developed a needle-type glucose sensor using a platinum electrode covered with immobilized glucose oxidase [178]. The current output was not significantly altered by changes in the oxygen tension of the solution in the range from 25–150 mm Hg. The glucose sensor was connected with a microcomputer system to calculate the insulin infusion rate and a syringe-driving system to deliver insulin. The sensor was inserted into the subcutaneous tissue of pancreatectomized dogs, and renewed every forth day. The artificial endocrine pancreas has kept effective in diabetic dogs with a glucose range of 5–9.5 mmol l^{-1} for 7 days.

In the past 50 years, clinical chemistry has developed increasingly sophisticated analytical methods, encompassing not only new tests for additional analytes but also new methodologies for recognized parameters. New methods such as UV and IR spectrophotometry, fluorometry, and HPLC have been introduced and the exploitation of enzyme and immunological techniques has yielded increasingly superior specificity in methods. No longer does the analyte have to be separated prior to its estimation; it can be measured directly in the tissue fluid. These measurements, however, use a bewildering assortment of complex chemistry and equipment.

Biosensors have considerable advantages over conventional techniques. Serum or blood are the most common fluids tested and the first biosensor will probably be geared to the usage of these samples [179]. A bench-top type of glucose analyzer has been commercially available and is being used in clinical chemistry laboratories. The blood or serum sample is injected onto a membrane impregnated with glucose oxidase, and the concentration of glucose is read out within 40 s. The instrument is automatically flushed and calibrated at the push of a button.

For fermentation applications, however, only a few glucose sensors have been investigated. ENFORS classified the sensors into two categories [180]:

(1) The analytical enzyme reactor in which the sample is withdrawn from the medium and pumped to the sensor site which is outside the fermentation vessel. The sample can be diluted or treated in different ways prior to contact with the sensor.
(2) The enzyme electrode, which can be immersed in the sample with the advantage of allowing measurements in situ provided that the sample is sterilizable. Since the sample is not diluted, the enzyme electrode is severely restricted in its upper linear measuring range which is determined by intrinsic enzymatic properties, i.e., the apparent Michaelis K_m of the immobilized enzyme preparation.

ENFORS proposed an externally buffered enzyme electrode, in which some of the advantages of the two types are combined [180]. The externally buffered enzyme electrode incorporates a flow-through system so that the enzyme chamber is continuously washed with a buffer solution.

Varied Enzyme Electrode. In a similar manner to the glucose sensor, a number of enzyme electrodes have been developed for determination of various organic substrates. Two types of electrochemical measuring devices are combined with matrix-bound enzymes:

(1) Amperometric devices: O_2 electrode, H_2O_2 electrode, H_2 electrode
(2) Potentiometric devices: H^+ electrode, CO_2 electrode, NH_3 electrode

Current measurement at a controlled potential is a feature of the amperometric device. The potentiometric device is characterized by potential measurement. Table 1.14 lists typical enzyme electrodes for organic molecules.

Most enzyme electrodes listed in Table 1.14 utilize single enzyme systems. Coupled bi- and multi-enzyme systems cover a broader range of substrates. Several multi-enzyme electrodes have been developed:

(1) A sucrose sensor using a multi-enzyme system of invertase, mutarotase, and glucose oxidase. Cholesterol esterase and cholesterol oxidase are incorporated in a cholesterol sensor.

Table 1.14. Enzyme sensors for organic molecules

Determinant		Biosensor	Biosensor assemblies	
			Receptor	Transducer
Saccharide	Glucose	Enzyme sensor	GOD	O_2, H_2O_2 electrode
		Enzyme thermistor	GOD/Catalase	Thermistor
		Microbial sensor	*P. fluorescens*	O_2 electrode
Alcohol	Ethanol	Enzyme sensor	Ethanol oxidase	O_2 electrode
		Microbial sensor	*T. Brassicae*	O_2 electrode
Amino acid	Glutamate	Enzyme sensor	Glutamate DH	NH_4^+ electrode
		Microbial sensor	*E. coli*	CO_2 electrode
Acid	Acetic acid	Enzyme sensor	Alcohol oxidase	O_2 electrode
		Microbial sensor	*T. brassicae*	O_2 electrode
	Uricate	Enzyme sensor	Uricase	O_2 electrode
		Enzyme thermistor	Uricase	Thermistor
Lipid	Cholesterol	Enzyme sensor	Colesterol esterase	Pt electrode
			Cholesterol oxidase	H_2O_2 electrode
		Enzyme thermistor	Cholesterol oxidase	Thermistor
	Phosphaty-dilcholine	Enzyme sensor	Phospholipase	H_2O_2 electrode
			Choline oxidase	
Antibiotics	Penicillin	Enzyme sensor	Penicillinase	H^+ electrode
		Enzyme thermistor	Penicillinase	Thermistor
	Cepharo-spollin	Microbial sensor	*C. freundii*	H^+ electrode
		Enzyme thermistor	Cepharospollinase	Thermistor
Urea		Enzyme sensor	Urease	H^+, NH_3, CO_2 electrode
		Enzyme thermistor	Urease	Thermistor
ATP		Enzyme thermistor	Hexokinase	Thermistor
Biotin		Bioaffinity sensor	Avidin/HABA	O_2 electrode
T_4		Bioaffinity sensor	Anti T_4/T_4	O_2 electrode
Mutagens		Microbial sensor	*B. subtilis* Rec	O_2 electrode
BOD		Microbial sensor	*T. cutaneum*	O_2 electrode

(2) RENNENBERG et al. reported enzyme electrodes consisting of several membranes coupled with an amperometric device [181]. Figure 1.40 illustrates the principle of the glucose dehydrogenase/glucose oxidase competition electrode for determination of NAD^+.

(3) KARUBE et al. have developed a novel sensor for freshness of fish [182]. Immobilized 5′-nucleotidase, nucleotide phosphorylase, and xanthine oxidase are combined with an oxygen electrode. Hypoxanthine (Hx), IMP, and inosine (HxP) are determined in sequence. The freshness index (K_I) is calculated by these parameters as follows:

$$K_I = (Hx + HxP)/(IMP + Hx + HxP) \times 100. \tag{1.27}$$

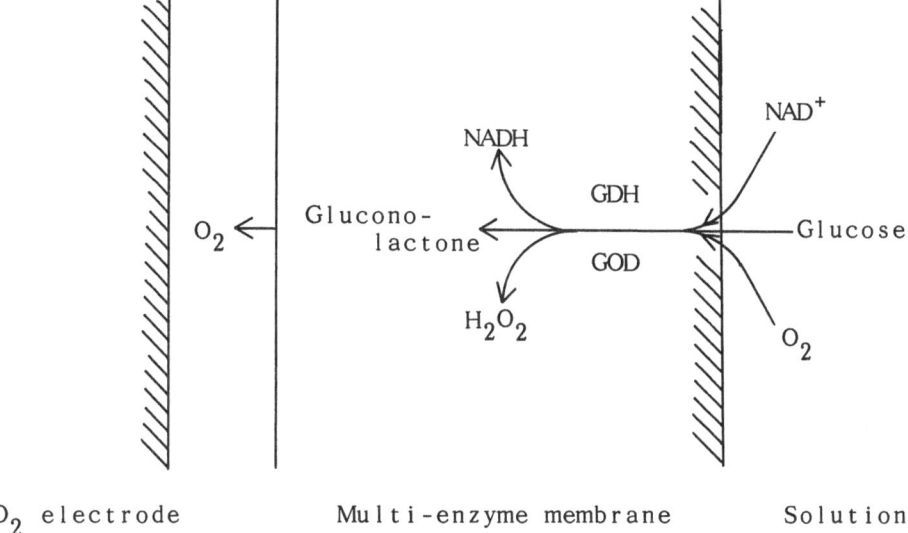

O_2 electrode   Multi-enzyme membrane   Solution

GDH : glucose dehydrogenase

GOD : glucose oxidase

Fig. 1.40. NAD^+ sensor with a multi-enzyme membrane

(4) Osawa et al. hybridized enzymes and microbial cells in a membrane-bound form to construct a sensor for creatinine [183]:

$$\text{Creatinine} \;-\!(\text{creatininase})\!\rightarrow\; \text{N-methylhydantoin} + NH_3, \tag{1.28}$$

$$NH_3 \;-\!(\textit{Nitrosomonas}\,\text{sp.})\!\rightarrow\; NO_2, \tag{1.29}$$

$$(\textit{Nitrobacter}\,\text{sp.})\!\rightarrow\; NO_3. \tag{1.30}$$

Sensitization of Enzyme Electrodes. Most enzyme electrodes may cover the analyte concentration range 10^{-5}–10^{-3} mol l^{-1}. The lower limit of detection has been improved for trace analyses [184, 185]. Mizutani et al. used enzyme cycling to enhance the sensitivity [186]. Enzyme cycling of lactate oxidase (LOD) and lactate dehydrogenase (LDH) was utilized as shown in Fig. 1.41 to detect lactate up to 10^{-8} mol l^{-1}.

Enzyme Electrodes for Enzyme Activity. Enzyme electrodes are sometimes mistaken for electrodes which measure enzyme activities. There are electrode systems which are specifically designed to assay for enzyme activity such as the use of the NH_3 gas electrode for kinetic determination of trypsin activity based on the hydrolysis of a synthetic substrate [187]. However, most enzyme electrodes couple

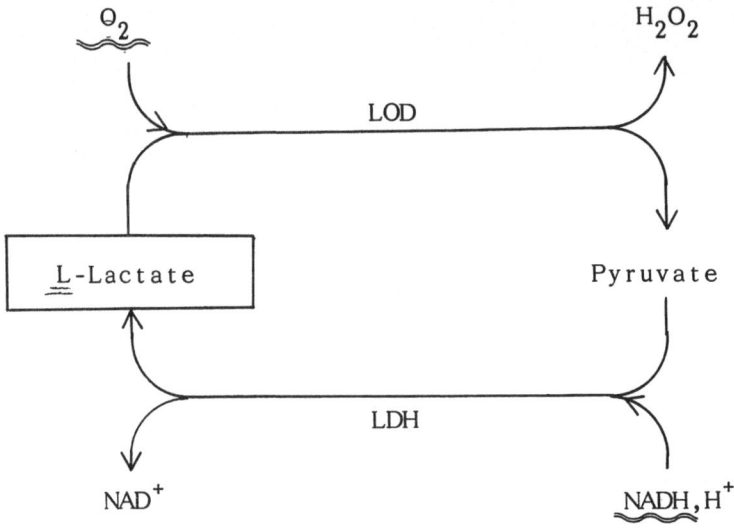

LOD : lactate oxidase

LDH : lactate dehydrogenase

Fig. 1.41. Enzyme cycle to improve the sensitivity of an enzyme electrode for lactate

an amperometric device with a specific enzyme that acts upon selected substrates as listed in Table 1.15 [188–191].

An enzyme electrode for pyruvate, which consists of membrane-bound pyruvate oxidase and an oxygen electrode, was coupled with immobilized oxaloacetate decarboxylase to determine glutamate pyruvate transaminase (GPT) and glutamate oxaloacetate transaminase (GOT) activities [188, 189]:

$$\text{2-Oxoglutarate} + \text{L-Alanine} \xrightarrow{\text{(GPT)}} \text{L-Glutamate} + \text{Pyruvate}, \quad (1.31)$$

$$\text{2-Oxoglutarate} + \text{L-Aspartate} \xrightarrow{\text{(GOT)}} \text{L-Glutamate} + \text{Oxaloacetate},$$
$$(1.32)$$

$$\text{Oxaloacetate} \xrightarrow{\text{(OAC)}} \text{Pyruvate} + \text{CO}_2, \quad (1.33)$$

$$\text{Pyruvate} + \text{O}_2 + \text{Pi} \xrightarrow{\text{(POP)}} \text{Acetylphosphate} + \text{CO}_2. \quad (1.34)$$

B. Enzyme Thermistors

Enzyme thermistors possess unique universality, since most enzyme reactions are accompanied by considerable heat evolution in the range of 5100 KJ/mol. An enzyme thermistor is a simplified flow calorimeter designed for routine analysis and is based on the use of immobilized enzymes. A preparation of immobilized en-

Table 1.15. Enzyme sensors for enzyme activity measurement

Determinant	Enzyme electrode	Reaction
Choline esterase	Choline oxidase	Acetylcholine + H_2O → Choline + CH_3COOH Choline + $2O_2$ → Betaine + $2H_2O_2$
LDH	Lactate oxidase	Pyruvate + NADH → Lactate + NAD^+ Lactate + O_2 → Pyruvate + H_2O_2
Alkaline phosphate	Glucose oxidase	G-6-P + H_2O → Glucose + Pi Glucose + O_2 → Gluconolactone + H_2O_2
Pyruvate kinase	Pyruvate kinase Pyruvate oxidase	ATP + Creatine → ADP + Creatinephosphate ADP + Phospho-enol-pyruvate → ATP + Pyruvate Pyruvate + Pi + H_2O + O_2 → CO_2 + H_2O_2
α-Amylase	Glucoamylase Glucose oxidase	Dextrin + H_2O → Glucose Glucose + O_2 → Gluconolactone + H_2O_2
GPT	Pyruvate oxidase	

Table 1.16. Enzyme thermistors

Determinant	Enzyme	Range ($mol \cdot l^{-1}$)
Saccharides		
Glucose	Glucose oxidase/catalase	$8 \times 10^{-4} – 2 \times 10^{-6}$
	Hexokinase	$2.5 \times 10^{-2} – 5 \times 10^{-4}$
Cellobiose	β-Glucosidase/glucose oxidase	$5 \times 10^{-3} – 5 \times 10^{-5}$
Lactose	Lactase/glucose oxidase	$1 \times 10^{-2} – 5 \times 10^{-5}$
Acids		
Ascorbate	Ascorbate oxidase	$6 \times 10^{-3} – 5 \times 10^{-5}$
Uricate	Uricase	$4 \times 10^{-3} – 5 \times 10^{-4}$
Lipids		
Cholesterol	Cholesterol oxidase	$1.5 \times 10^{-4} – 3 \times 10^{-5}$
Cholesterol ester	Cholesterol oxidase/esterase	$1.5 \times 10^{-4} – 3 \times 10^{-5}$
Triglyceride	Lipoprotein lipase	$5 \times 10^{-3} – 1 \times 10^{-4}$
Antibiotics		
Cepharospollin	Cepharospollinase	$1 \times 10^{-2} – 5 \times 10^{-6}$
Pennycillin G	Pennycillinase	$5 \times 10^{-1} – 1 \times 10^{-5}$
Miscellaneous		
ATP	Hexokinase	$8 \times 10^{-3} – 1 \times 10^{-3}$
Urea	Urease	$5 \times 10^{-1} – 1 \times 10^{-5}$
Heavy metal ion	Urease	10^{-9}
Pesticide	Acetylcholine esterase	5×10^{-6}

zyme is placed in close contact with a thermistor. When substrate is fed to the enzyme, product is formed together with the liberation of a small amount of heat. This heat effect, measured as a temperature increase by the thermistor is used to detect and register the concentration of the substrate.

A number of enzyme thermistors have been reported, primarily by MOSBACH et al. [192, 193]. Practically useful enzyme thermistors have been designed for the

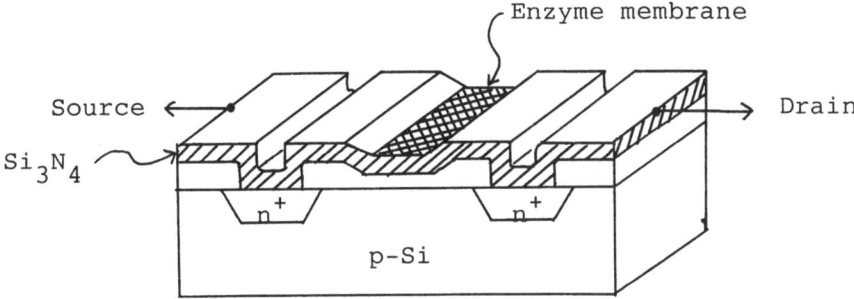

Fig. 1.42. Configuration of enzyme FETs

determination of urea, glucose, cholesterol, and cholesterol esters, sugars, phenol, and hydrogen peroxide (Table 1.16). The applicability and properties of enzyme thermistors were first investigated in the field of clinical chemistry and environmental control analysis. Further application has been presented in which enzyme thermistors have been used for process control.

C. Enzyme FET Sensors

A family of new biosensors has been studied in the last decade using the MOSFET structure. The new biosensors consist of immobilized enzyme attached to ion-sensitive field effect transistor (ISFET) which is essentially an insulated gate field transistor without a metal gate. The operation of the ISFET is similar to that of the insulated gate FET in that the reference electrode and the electrolyte constitute the modified gate. The interfacial potential at the electrolyte/insulator interface produced by the ions in solution will affect the channel conductance of the ISFET in the same way as the external gate voltage applied to the reference electrode. Several ISFETs have been developed for the determination of pH, Na^+, K^+, and Ca^{2+} [194–196].

JANATA et al. proposed an enzyme FET sensitive to penicillin [197]. The device was constructed by depositing a co-cross-linked penicillinase albumin layer over a pH sensitive FET. Immobilized penicillinase recognizes penicillin in solution with a resulting change in pH, which is detected by the FET. MORIIZUMI and KURIYAMA et al. have developed improved enzyme FETs sensitive to urea and glucose [198, 199]. The configuration of these enzyme FETs is illustrated in Fig. 1.42.

LUNDSTROM et al. showed that the $Pd-SiO_2–Si$ structure of MOSFET can be used for monitoring hydrogen dissolved in electrolytes [200]. It has been found that the NH_3 sensitivity is enhanced by depositing small amounts of other catalytic metals such as Ir on the Pd film. These Pd-MOSFETs were used to form a new type of biosensor, the enzyme transistor [201]. The enzyme transistor differs from the enzyme FET in that the latter device is based on ISFET, while the enzyme transistor is based on gas-sensitive FET.

The advantages of these bioelectronic sensors, including enzyme FETs and enzyme transistors, are the possibilities of minituarization and direct integration with microelectronics. It should even be possible to develop multichannel devices

with several different enzymes combined with different sensitive areas on the chip.

D. Enzyme Photodiode Sensors

A photodiode or phototransistor is coupled with immobilized enzyme which catalyzes a luminescent reactor [202]. The device is termed enzyme photodiode (or transistor). Peroxidase catalyzes the following reaction:

$$\text{Luminol} + H_2O_2 \rightarrow \text{Aminophthalate} + N_2 + h\nu. \tag{1.35}$$

The concentration of hydrogen peroxide is determined by counting photons at a constant concentration of luminol. Peroxidase was immobilized onto a Si photodiode to form an enzyme photodiode sensitive to hydrogen peroxide. The conceptual scheme of the device is illustrated in Fig. 1.43.

A glucose sensor was constructed by incorporating glucose oxidase together with peroxidase onto a Si photodiode.

E. Optrode

The term optrode, formed by combining "optical" and "electrode" emphasizes that the use of optical sensors is very similar to that of electrodes [203]. The de-

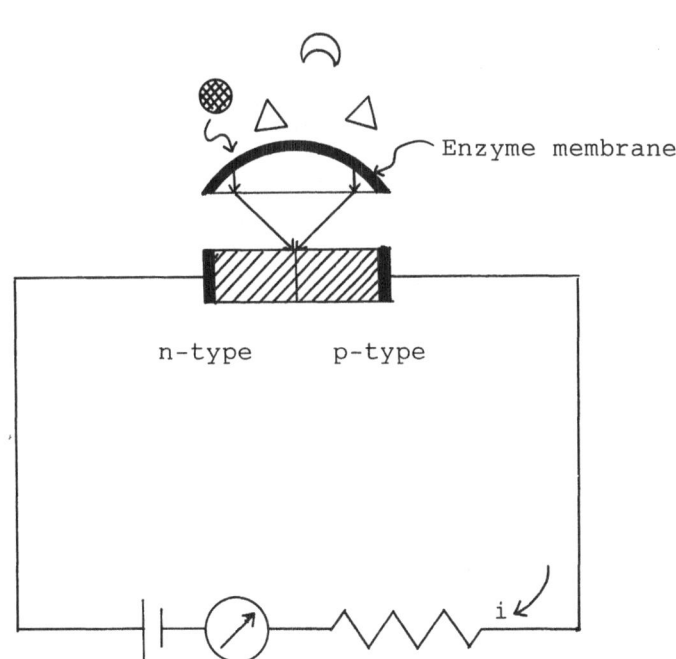

Fig. 1.43. Enzyme photodiode

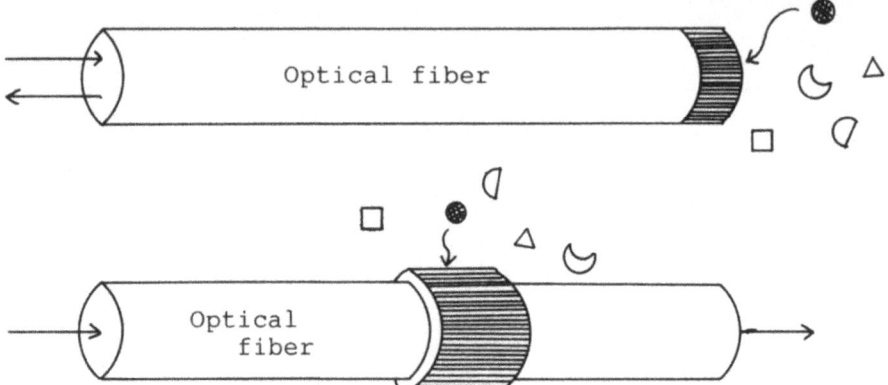

Fig. 1.44. Optical biosensor

vices involve a molecular recognition part on the end of a fiber optic. In operation, interactions with the determinant lead to a change in optical properties of the molecular recognition part, which is probed and detected through the fiber optic. Depending on the particular device, the optical property measured can be absorbance, reflectance, luminescence, or something else (Fig. 1.44).

SEITZ et al. reported optrodes sensitive to pH [204]. A more recent pH sensor based on fluorescence illustrates two wavelength measurements. An important example of a sensor based on quenching is the oxygen probe developed by PETERSON et al. [205].

A fluorescence-based, competitive biosensor for glucose has been described. The specific glucose-binding reagent is concanavalin A immobilized on Sepharose. The competing ligand is dextran labeled with fluorescein.

AIZAWA et al. utilized enzymatic luminescence to detect hydrogen peroxide and glucose [206]. Furthermore, optical sensors coupled with immobilized antibody were developed for the determination of antigen, in which luminescent enzyme was used as a label [207]. It was shown that hemin worked as a label in the optical immunosensor [208].

1.5 Bioelectrochemical Energy Conversion

Bioelectrochemical energy conversion is based on the utilization of biomolecules and microorganisms in conjunction with electrochemical systems. This section deals with bioelectrochemical photoenergy conversion and biochemical fuel cells that are two major energy conversion systems involved in this category (Fig. 1.45).

In bioelectrochemical photoenergy conversion it is attempted to mimic the photosynthetic process that takes place in the thylakoid membrane of the chloroplast. Many photoelectrochemical cells incorporated into photosynthetic pig-

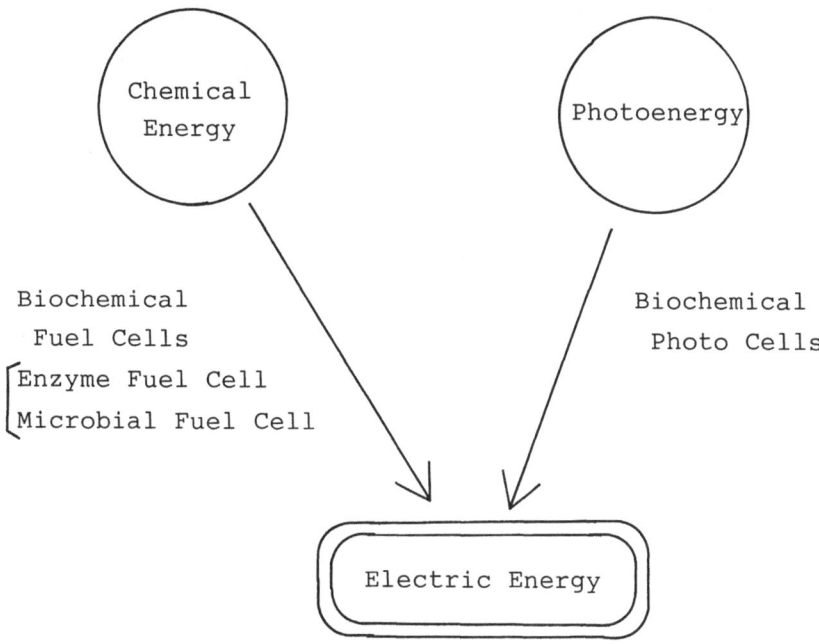

Fig. 1.45. Bioelectrochemical energy conversion

ments have been intensively investigated. Whole chloroplasts and living cells also have been incorporated into electrochemical systems to convert photoenergy.

The biochemical fuel cell can generate electricity by the electrochemical oxidation of organic chemicals (fuels) which are less electroactive. Many biochemical reactions catalyzed by enzymes and microorganisms are incorporated into biochemical fuel cells to make these organic chemicals electroactive in various manners. Such a fuel cell can hardly be accomplished simply by utilizing electrochemical reactions. Although the biochemical fuel cells are classified into enzyme fuel cells and microbial fuel cells depending on what type of biocatalyst is used, both fuel cells operate by the same principle.

1.5.1 Bioelectrochemical Photoenergy Conversion

1.5.1.1 Architecture of Energy-Transducing Biomembrane

Biophysical-chemical studies have suggested that the thylakoid membrane of the chloroplast consists of an ultrathin layer of lipids (presumably a lipid bilayer) with sorbed proteins and pigments organized in a lamellar structure approximately 100 Å thick. Although the precise functions of the thylakoid membrane are still obscure, it is believed that the membrane is the locus of the primary photophysical and photochemical processes.

During the last several years, purple membranes formed from the extreme halophile *Halobacterium halobium* and its retinal-containing protein, bacteriorhodopsin, have attracted the attention of many groups working in the field of bioenergetics. Unlike photosynthesis and respiration, the function of bacteriorhodopsin does not rely on the transport of electrons but instead relies on the vectorial transport of proteins which leads to the establishment of an electrochemical proton gradient across the cell membrane. The energy for proton translocation is provided by light which is absorbed by the chromophore of bacteriorhodopsin.

1.5.1.2 Molecular Organization of Photoenergy-Transducing Synthetic Membranes

Various types of chlorophyll-containing membranes have been designed for photoenergy conversion primarily on the basis of the idea that photosynthesis is an electronic process. The process assumes that two-dimensional chlorophyll arrays are photoconductive. The absorption of a photon promotes one electron into the conduction band and leaves one hole in the valence band. The excited electrons and holes are free to move around. They are also considered to be spread out over the whole unit although their lifetimes and mobilities may be different. In the second scheme, it is assumed that the electrons can be transferred to an electron acceptor which is thereby effectively reduced, and the holes can be transferred to an electron donor which is thereby effectively oxidized.

Photosynthetic pigments, specifically chlorophylls, have been attached to a variety of matrices to convert visible light energy to electric or chemical energy. The molecular assemblies may be classified into the following three categories.

Pigmented Bilayer Lipid Membrane. Chlorophyll molecules have been incorporated into two types of artificial bilayer lipid membrane systems for the study of photoenergy transduction. The first consists of a planar bilayer lipid membrane (BLM) separating two aqueous solutions where photovoltaic effects can be induced. The second system comprises liposomes which are ideally suited for studies of photo-induced permeability, spectroscopy, and chemical reactions. For more complete technical details, two pertinent publications are available [209, 210].

BLMs are made of pure lipid, for example, phosphatidyl choline, or oxidized cholesterol in common salt solutions. Chlorophyll molecules situated at the biface in the Chl-BLM are tightly compressed together, and the average area occupied per porphyrin group is much less than 75 Å^2 [210, 211]. It is assumed that the hydrophobic protons (phytyl group) of the molecules extend inwards, while the hydrophilic heads (porphyrin) and polar lipid groups are situated at the aqueous solution/membrane interface. On the basis of this data, it seems probable that the porphyrin plane of the chlorophyll molecule is oriented at about 45° to the lipid bilayer [212].

Chlorophyll-Containing Membrane on an Electrode Surface. Photoelectrochemical and photovoltaic effects have been noted with a variety of chlorophyll assemblies deposited on metal or semiconductor electrode surfaces.

In the photosynthetic primary processes, electrons are pumped *via* two photosystems (photosystem I and II), where light-induced charge separation takes place efficiently in association with uni-directional electron transport. The oxidative water-splitting reaction is linked to photosystem II; photosystem I is followed by the reduction of NADP. The function of photosystem I can thus be simulated by a photocathode in an electrochemical photocell.

An n-type semiconductor electrode modified with a chlorophyll-containing membrane may act as an efficient photoanode for visible light conversion. TRIBUTSCH and CALVIN first employed chlorophylls and demonstrated an anodically sensitized photocurrent under potentiostatic conditions in aqueous electrolytes [213]. The photocurrent action spectrum of chlorophyll *a* on ZnO (single crystal) showed a red band peak at 673 nm, corresponding closely to the amorphous and monomeric state of chlorophyll *a*. The anodic photocurrent increased by the addition of supersensitizers (reducing agents) to the electrolyte. A maximum quantum efficiency of 12.5% was obtained in the presence of phenylhydrazine. TAKAHASHI et al. made a thin membrane of chlorophylls complexed with electron donors which was deposited on a platinum electrode [214, 215]. It should be noted that the chlorophyll electrode behaved as a photocathode when electron acceptors were co-deposited in place of the electron donors. The photoactive species could be attributed to the composite of chlorophyll-oxidant or chlorophyll-reductant.

To enhance the quantum efficiency, various molecular structures have been proposed for photoenergy transduction using chlorophyll-containing membranes. FONG et al. [216] insisted that hydrated microcrystalline chlorophyll *a* is superior to the amorphous aggregate in light-induced charge separation. Hydrated microcrystalline chlorophyll *a*, the dehydrated Chl *a* oligomers, (Chl a-$2H_2O)_n$, are considered to be the most powerful redox species for water splitting under far red excitation. The water-splitting reaction was demonstrated with an illuminated (Chl a-$2H_2O)_n$-platinum electrode [217]. The authors indicated that the (Chl a-$2H_2O)_n$-platinum electrode is capable of reducing CO_2 under illumination.

AIZAWA et al. [218–221] incorporated chlorophyll molecules into a liquid crystal thin membrane, which was deposited on a platinum electrode. The liquid crystal molecules were found to effectively inhibit the intermolecular interaction of chlorophylls and the formation of the chlorophyll hydrate, which could enhance the charge separation in photoexcitation.

Attempts have been made to intersperse a suitable inert surfactant diluent in the monolayer in order to control the surface concentration as well as the mean intermolecular separation of chlorophyll. Among the surfactants which can act as ideal two-dimensional diluents for a chlorophyll *a* monolayer, MIYASAKA et al. [222] chose the phospholipid dipalmitoyllecithin (DPL) to enhance the quantum efficiency. They conclude that a monolayer thick membrane having well-interspersed chlorophyll is most efficient for the conversion of light energy using chlorophyll-sensitized semiconductor electrodes. AIZAWA et al. [223] introduced various substances among chlorophyll molecules to prevent the intermolecular interaction and to improve the photoconversion properties. A monolayer of Cu-chlorophyllin was fixed on the surface of SnO_2 *via* heavy metal polynuclear chelation.

The monolayer membrane was then partially substituted by Cr-myristate. It should be noted that the quantum efficiency was sharply increased by the presence of Cr myristate. A thin layer of chitosan, in which Cu-chlorophyllin was immobilized, was deposited on the surface of SnO_2. The chitosan markedly enhanced the quantum efficiency in a similar manner to myristate. Chitosan polymer may be considered to effectively prevent the energy dissipation of surface-bound sensitizer in photoexcitation.

Bacteriorhodopsin-Containing Membrane. The preparation of BLM containing bacteriorhodopsin was first reported by DANCSHAZY and KARVALY [224] and HERRMANN and RAYFIELD [225]. These membranes developed a maximum photocurrent of 2×10^{-12} A and a photopotential of 20–60 mV, the action spectrum of which indicated that bacteriorhodopsin was the species responsible for the photoelectric effect. SHIEH and PACKER [226] increased the lifetime of the film by adding of polystyrene or polyacrylamide to the membrane forming solution. Efforts to increase the total surface area of the artificial membrane have been most successful by attaching liposomes containing bacteriorhodopsin to Millipore filters [227]. PACKER et al. [228] reported that the membrane preparations retained their photovoltaic response for over a month without significant loss of activity.

1.5.1.3 Methodology of Energy-Transducing Membrane Formation

Pigmented BLM. Planar bilayer lipid membranes (BLM) consist of a variety of lipids of highly purified single phospholipids. BLM are commonly formed in an aperture in a hydrophobic barrier (for example, a Teflon cup) separating two aqueous solutions. A droplet of lipid solution is applied to the aperture. Excess lipid solution drains into the border and within minutes optically black spots appear and enlarge themselves to occupy the majority of the aperture area. Pigmented BLM can be formed from a droplet of lipid solution containing pigment such as chlorophyll or bacteriorhodopsin.

Because BLM made of pure lipid or oxidized cholesterol in common salt solution are nonconducting, the physical properties of BLM are with one exception similar to those of a lipid hydrocarbon layer of equivalent thickness. The interfacial tension of BLM is less than 5 dynes cm^{-1}, which is approximately one order of magnitude lower than that of the hydrocarbon/water interface. This low interfacial tension is due to the presence of polar groups at the interface. BLM have negligible permeability for ions and most polar molecules. Permeability to water is comparable to that of biological membranes. The permeability to water in chlorophyll BLM, as determined by an osmotic flow method, is 50 μm^{-1}, which is in the range of phospholipid BLM but six times larger than that of oxidized cholesterol BLM.

Pigment Membrane Deposited on a Solid Surface. An amorphous membrane is generally prepared by solvent evaporation of an organic solution of chlorophyll or bacteriorhodopsin on a solid substrate surface. The red absorption peak of a dry amorphous chlorophyll *a* membrane is approximately 675–680 nm. This reflects the formation of dimers and oligomers of chlorophyll species which are readily obtained by allowing a small amount of water to remain in a solution of chlorophyll *a* dissolved in a suitable nonpolar organic solvent. Membranes can be formed on a solid surface by solvent evaporation. A chlorophyll *a*-H_2O microcrystalline membrane also has been prepared by the electrodeposition technique established by TANG and ALBRECHT [229].

Pigmented Monolayer Membrane on the Solid Surface. Monolayer and mixed monolayers of pigments can be prepared on a neutral aqueous buffer solution surface and deposited, at a controlled surface concentration, onto a solid substrate by the Langmuir-Blodgett technique [230, 231]. As a typical feature, a monolayer membrane of chlorophyll *a* possesses absorption peaks at 675–680 nm and 435–440 nm with corresponding optical sensitivities of 0.008–0.01 and 0.01–0.013 in the red and blue bands, respectively [232].

1.5.1.4 Photoenergy Transduction

Pigmented BLM elicits large photopotentials under asymmetric conditions, particularly when the membrane is interposed between two solutions containing different redox couples. For example, the presence of $FeCl_3$ on one side and ascorbic acid on the other side resulted in a photopotential of more than 150 mV across a chlorophyll-BLM [233]. A variety of redox compounds affecting the photoresponse of the chlorophyll-BLM have been investigated [234, 235].

Bilayer lipid membranes containing chloroplast extracts are almost ideal systems for investigating energy transduction and photochemical processes as they occur in the photosynthetic membranes of green plants. Although many of the obtained results are qualitative and preliminary, the pigmented BLM have opened a new vista to the study of the intricate membranous aspects of photosynthesis, particularly the primary events of solar energy conversion [234, 235]. GROSS et al. [236] made a photovoltaic cell using photosystem I subchloroplast particles (Fig. 1.46). The particles were placed on a filter between two compartments, one of which contained the electron donor $K_4Fe(CN)_6$ and the other the electron acceptor FMN. Upon illumination with white light ($I = 80$ Wm^{-2}), a potential of 300 mV generated across a 3000 Ω load resistance. Both PS I photochemistry and direct photoreactions of FMN contributed to the process. A power output of 20 W was observed for a 2 cm^2 filter containing 60 µg chlorophyll. This corresponds to 0.1 Wm^{-2}. The power efficiency was 0.13%. The short circuit current was 108 µA [236]. KATZ et al. reported a photovoltaic cell comprised of a glass cylinder containing two compartments separated by either a polyvinyl chloride membrane or aluminum foil. A chlorophyll-water adduct was impregnated in the membrane or deposited on one side or the Al foil by evaporation from an octane suspension of the adduct. One of the compartments in the cylinder was filled with an electron acceptor solution such as tetramethylphenylenediamine. The second compartment was filled with an electron donor solution such as sodium ascorbate. Each compartment contained an electrode. In the best cell version – using an albumin film support for the chlorophyll – a potential difference of 422 mV was observed with a simultaneous current of 2.36×10^{-5} A in the external circuit. The efficiency was estimated at about 0.0024%. TIEN et al.

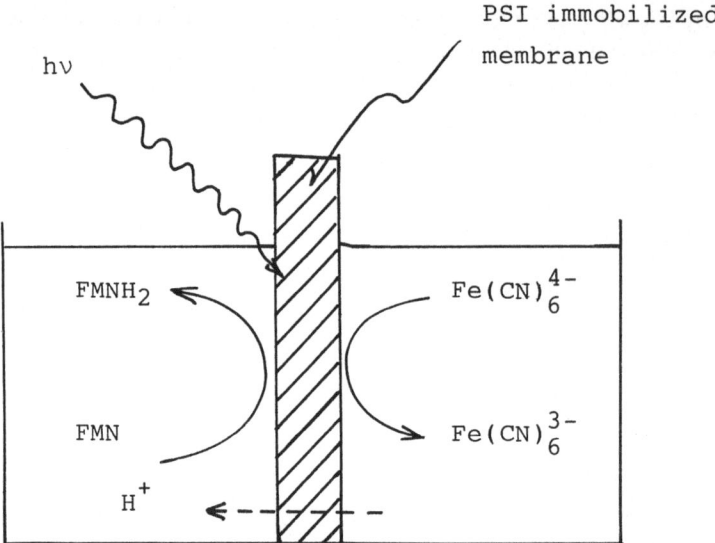

Fig. 1.46. Light-driven uphill transport of electrons across a PSI-immobilized membrane

incorporated chlorophyll-BLM (phosphatidyl choline) into a porous polycarbonate filter (pore size 0.05 to 8 μm) to make a photovoltaic cell.

OCHIAI et al. prepared membrane-bound whole chloroplasts [237]. A chloroplast suspension was mixed with native ferredoxin and co-immobilized with polyvinyl alcohol (PVA-177). The resulting film reduced NADP upon illumination at 0.64 mole per mg chlorophyll per h, with about 1% yield of recovery. They applied the film to a photovoltaic cell.

Photoelectrochemical conversion from visible light to electric and/or chemical energy has been investigated with chlorophyll thin membranes deposited on semiconductors or metal electrodes [238]. Chlorophyll-coated metal (platinum) electrodes yielded a cathodic photocurrent in acidic electrolyte solutions, although the photocurrent efficiencies tend to be low compared to those of chlorophyll-semiconductor electrodes. The cathodic photoresponse may result from the p-type photoconductive nature of a solid chlorophyll layer and/or the formation of a contact barrier at the metal-chlorophyll interface, which contributes to light-induced carrier separation and leads to photocurrent generation.

A photoelectrochemical cell reported by TAKAHASHI et al. contained chlorophyll-naphthoquinone-Pt and chlorophyll-anthrahydroquinone-Pt electrodes [214]. In a typical experiment, NAD and $Fe(CN)_6^{-4}$, each dissolved in neutral electrolyte solution, were employed as an acceptor for a photoanode. Upon illumination with white light, the short circuit current was $8 \times 10^{-6} \, A \cdot cm^{-2}$.

AIZAWA et al. showed that chlorophyll-liquid crystal electrodes in acidic buffer solutions gave cathodic photocurrents accompanied by the evolution of hydrogen gas [219]. Substitution of the central metal of native chlorophyll (Mg-pheophytin) resulted in a drastic change of photoelectrochemical behavior. A Mn-pheophytin/liquid crystal/Pt electrode generated an anodic photocurrent upon illumination.

In contrast, a Ru-pheophytin/liquid crystal/Pt electrode gave a cathodic photocurrent with a quantum efficiency of approximately 0.5% [221]. FONG'S work on photoelectrochemical cells has proved that chlorophyll hydrates are photoactive [216].

Chlorophyll-coated semiconductor (n-type ZnO, CdS, and SnO_2) electrodes provided dye-sensitized anodic photocurrents with high efficiency. It was found that the quantum efficiency of the anodic photocurrent in the chlorophyll a/dipalmitoyl lecithin (DPL)SnO_2 system was enhanced to about 25%, while the chlorophyll a monolayer/SnO_2 gave a quantum efficiency of 3–4%. The considerable increase in quantum efficiency may result from the suppression of the self-quenching of chlorophyll molecules [222]. MIYASAKA et al. [222] concluded that a monolayer thick film having well interspersed chlorophyll is most efficient for the conversion of light energy using chlorophyll-sensitized semiconductor electrodes. AIZAWA et al. [223] also have shown that the use of potential barrier molecules between chlorophyll molecules may enhance the quantum efficiency.

1.5.1.5 Photoenergy Transduction with Chloroplast Containing Electrodes

Chlorophyll-containing membranes and electrodes have been assembled to mimic either the photosystem I or II. On the other hand, much effort has been concentrated on the immobilization of photosystems I and II, and of intact chloroplasts in membrane matrix attached to the electrode surface. The immobilized photosynthetic organella generate electroactive substances upon illumination, which results in a photoresponse of the electrode.

GROSS et al. immobilized photosystem I particles to a cellulose triacetate membrane with agar-agar [236]. A two-compartment electrolytic cell was separated with the photosystem I immobilized membrane. The structure is shown in Fig. 1.46. Photoexcited photosystem I accepts electrons from ferrocyanide in one compartment, which results in subsequent donation of electrons to FMN in the other compartment. This system showed a light-driven uphill transport of electrons.

IIDA et al. proposed a photoelectrochemical cell incorporating the chloroplast/methyl viologen system in the anolyte [239, 240]. Class II chloroplasts from spinach were suspended in an anolyte containing methyl viologen. The catholyte, separated from the anolyte with a salt bridge, contained a H_2SO_4 solution. A SnO_2 and a Pt electrode were used as anode and cathode, respectively. The chloroplasts extract electrons from H_2O upon illumination, followed by the reduction of methyl viologen. Electrochemical oxidation of reduced methyl viologen takes place at the anode. At the counterelectrode, hydrogen evolution may occur through electrochemical reduction of protons. The postulated reaction scheme is illustrated in Fig. 1.47.

OCHIAI et al. immobilized chloroplasts on the SnO_2 electrode surface [241]. The chloroplast electrode was shown to function as a photoanode. Addition of electron carriers such as 2,6-dichlorophenol indophenol to the anolyte resulted in an extreme increase of the photocurrent.

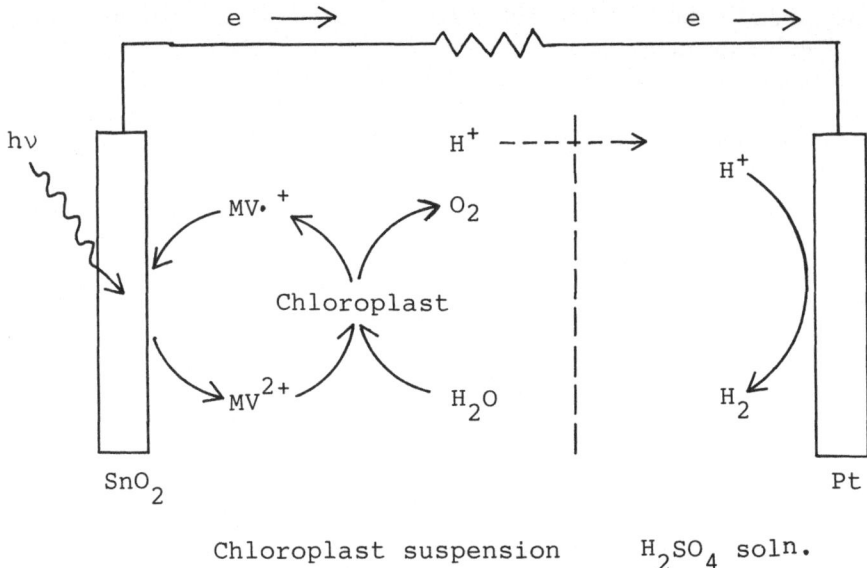

$$MV \cdot^+ \qquad O_2$$

Chloroplast

$$MV^{2+} \qquad H_2O$$

SnO$_2$ Pt

Chloroplast suspension   H$_2$SO$_4$ soln.

Fig. 1.47. Photoelectrochemical cell incorporating the chloroplast/methyl viologen system

It was shown that the intact cells of thermophilic *Mastigocladus laminosus*-MS retained the photosynthetic activity of photosystem I and II [242]. Whole cells of *Mastigocladus laminosus*-MS were immobilized with calcium alginate on the SnO$_2$ electrode surface. The modified electrode generated a photocurrent as in the case of chloroplast-bound electrodes. The photocurrent was profoundly enhanced by heat-treatment of the cell-immobilized electrode at 40–50 °C for 50 min. Although chloroplasts from spinach lose their photosynthetic activity of photosystem II at 45 °C, *M. laminosus* retains the photosynthetic activity of photosystem I and II at the same temperature.

1.5.1.6 Photo-Generation of Hydrogen Coupled with a Fuel Cell

Green algae, blue-green algae, and photosynthetic bacteria generate hydrogen upon visible light illumination. Several microorganisms capable of photo-generation of hydrogen are included in Table 1.17.

Since the photo-generation of hydrogen results from complicated reactions, the mechanisms which differ among these microorganisms have been uncertain. In green algae, however, hydrogenase should link the photosystem with hydrogen generation through ferredoxin. On the other hand, in photosynthetic bacteria and blue-green algae nitrogenase should take part in the reduction of protons to generate hydrogen.

Several systems for photo-generation of hydrogen have been assembled as follows:

Chloroplast/Hydrogenase System. Intact chloroplasts produce the reduced form of ferredoxin upon illumination. Hydrogenase can extract electrons from the re-

Table 1.17. Microorganisms capable of photo-generation of hydrogen

Green algae	*Scenedesmus* sp.
	Chlorella sp.
	Chlamydomonas sp.
	Ankistrodesmus sp.
Blue algae	*Anacystis nidulans* et al.
	Synechococcus sp.
	Anabaena cylindrica et al.
	Mastigocladus laminosus et al.
	Oscillatoria sp.
Photosynthetic bacteria	*Rhodospirillum rubrum*

duced form of ferredoxin to reduce proton to hydrogen. Therefore, this reaction is coupled with hydrogenase as shown in Fig. 1.48.

Arnon et al. reported in 1961 that hydrogen evolved in the system containing chloroplasts from spinach, hydrogenase and ferrodoxin from *Clostridium sp.,* and cysteine [243]. Oxygen evolution was not observed.

In 1973, Benemann constructed a similar system, except that it contained no cysteine, which generated hydrogen upon illumination with simultaneous oxygen evolution [244]. Oxygen evolution, however, severely damaged hydrogenase. Although addition of the glucose/glucose oxidase system prevents inactivation of hydrogenase, it should not be photodecomposed by water.

Yagi et al. proposed a unique photoelectrochemical system consisting of a chloroplast suspension in a catholyte compartment and a methyl viologen solution in an anolyte compartment [245, 246]. The two compartments were separated

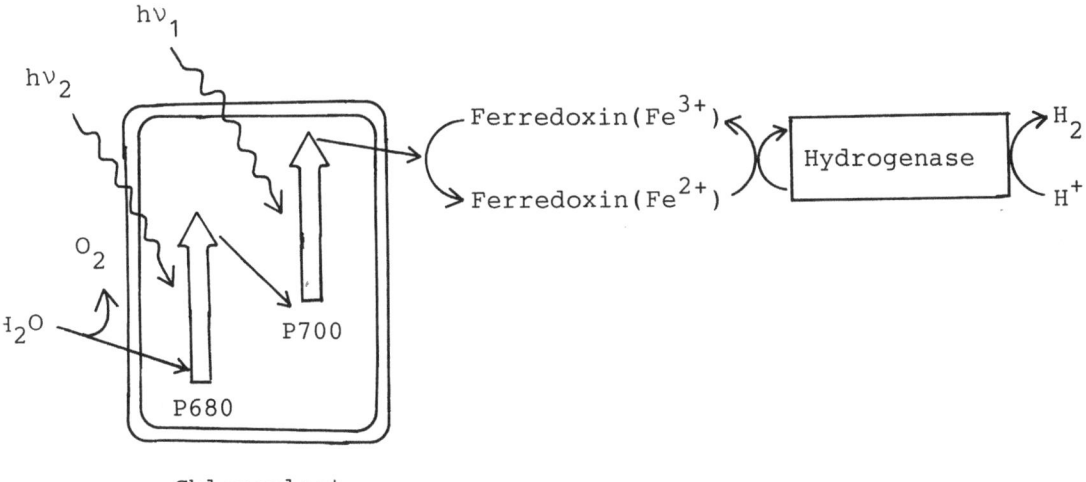

Fig. 1.48. Chloroplast/hydrogenase system for the photo-generation of hydrogen

by a salt bridge. Photoillumination of suspended chloroplasts caused an increase in the photocurrent. Addition of 1-methoxy PMS remarkably enhanced the photocurrent.

Immobilized Green Algae/Immobilized H-Producing Bacteria System. The chloroplast/hydrogenase system should suffer from instability of the photosynthetic molecular assemble and hydrogenase. In order to overcome the problem, chloroplasts and hydrogenase were immobilized onto solid matrices. KARUBE et al. entrapped chloroplasts from spinach in polyacrylamide gel, which retained the activity of NADH reduction for several hours [247, 248]. Hydrogenase-bearing *Cl. butyricum* were immobilized in agar gel while retaining their hydrogenase activity [249].

The immobilized chloroplast and hydrogenase were packed in a separate column to form reactors as shown in Fig. 1.49. These reactors were connected with a fuel cell. Ferredoxin was circulated through these reactors and the fuel cell. Immobilized chloroplasts reduce ferredoxin upon illumination. Reduced ferredoxin is transferred to the hydrogenase reactor, where hydrogen evolves by oxidation of reduced ferredoxin. Hydrogen is used to generate electricity in a fuel cell. The oxidized ferredoxin is returned to the photochemical reactor. Continuous illumi-

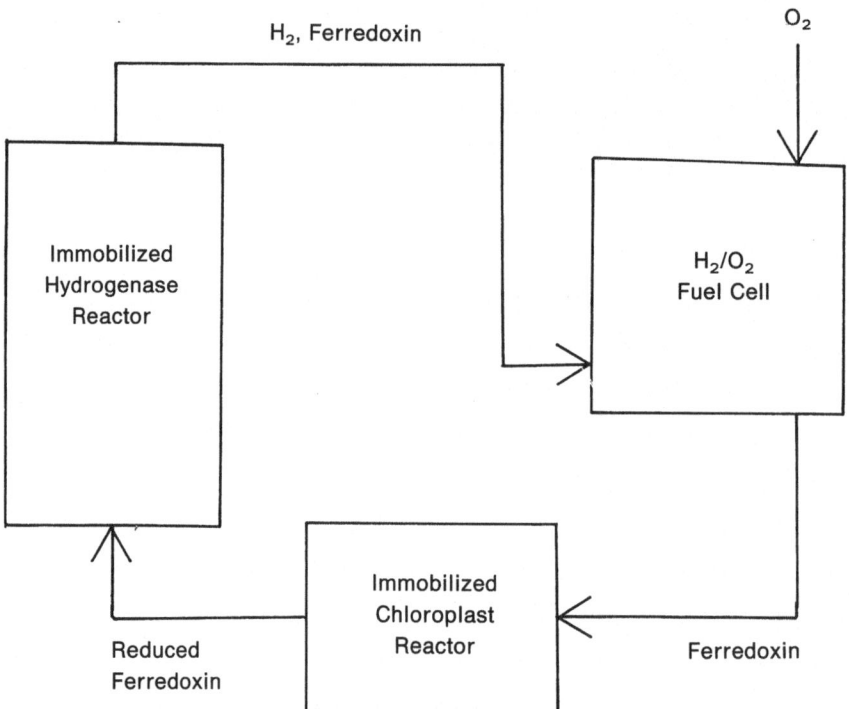

Fig. 1.49. Immobilized chloroplast/immobilized hydrogenase system for the photo-generation of hydrogen coupled with a hydrogen/oxygen fuel cell

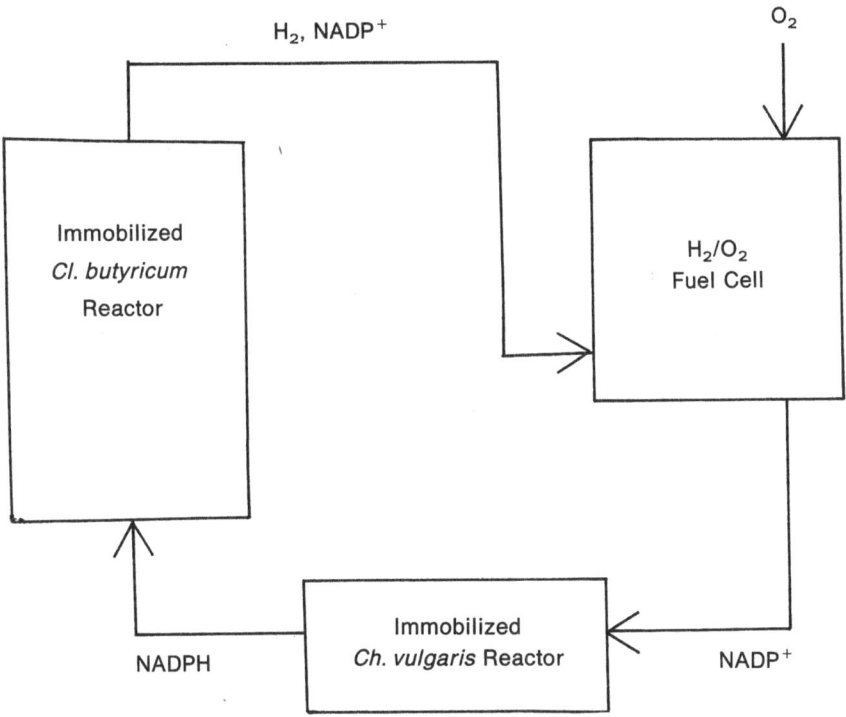

Fig. 1.50. Immobilized green algae/immobilized *Clostridium butyricum* for the photo-generation of hydrogen coupled with a hydrogen/oxygen fuel cell

nation with circulation of the inner solution generates a current. The current, however, gradually decreased due to the inactivation of immobilized chloroplasts.

An improved system has been developed for the photo-generation of hydrogen. Green algae such as *Chlorella vulgaris* were immobilized in a gel. The immobilized green algae provided a very stable activity of photo-reduction of NADP under anaerobic conditions. NADP was found to be a proper electron carrier to link the immobilized *Ch. vulgaris* with the immobilized *Cl. butyricum*. A photochemical reactor with immobilized *Ch. vulgaris* was connected to a reactor with immobilized *Cl. butyricum*. Hydrogen was continuously produced with circulation of NADP under illumination. The system for photo-generation of hydrogen was coupled with a hydrogen/oxygen fuel cell as shown in Fig. 1.50, which yielded a stable current for 6 days [250].

Immobilized Blue-Green Algae System. Anabaena sp. have a heterocyst structure and generate hydrogen from water upon illumination. Photo-generation of hydrogen is retarded by oxygen. KAYANO et al. utilized aerobic bacteria, *B. subtilis,* to consume oxygen generated by Anabaena [251]; the system is depicted in Fig. 1.51.

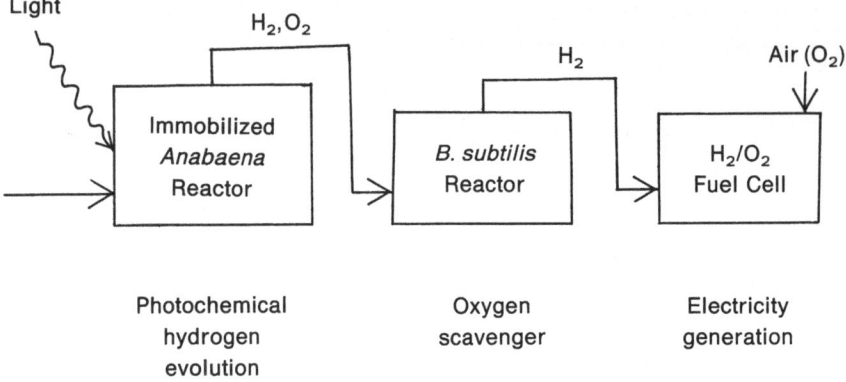

Fig. 1.51. Immobilized blue-green algae/aerobic bacteria system for the photo-generation of hydrogen coupled with a hydrogen/oxygen fuel cell

1.5.2 Enzyme Fuel Cell

Enzyme fuel cells may be classified into three basic types of cells: the product cell, the depolarizer cell, and the regenerative cell [252]. The fundamental principles of these fuel cells are schematically illustrated in Fig. 1.52.

In product cells, enzymes are used to convert compounds that are not electro-chemically active into electrochemically active ones. In depolarizer cells, enzymes act as depolarizers of electrochemical reactions such as the oxidation of H_2 or the reduction of O_2. In regenerative fuel cells, enzymes are used to regenerate redox compounds that take part in the electrochemical reactions.

Two reasons should be pointed out for investigating biological redox systems: to attempt to employ them directly, and to understand them sufficiently well to reproduce, model, or mimic the essential features in simpler compounds.

1.5.2.1 Historical Background

The conceptual root of enzyme fuel cells may be found with POTTER's experiment in 1911 to determine the activity of microorganisms by measuring the potential difference between a pair of electrodes in a yeast-glucose solution and a glucose solution [253]. A small current was obtained by COHEN in 1932 from a microbial fuel cell [254]. In 1962, ROHRBACK et al. utilized hydrogen-producing bacteria, *Cl. butyricum,* to construct a microbial fuel cell [255]. Glucose was converted to hydrogen by a biochemical reaction, which was followed by the electrochemical conversion of hydrogen to electricity. DELDUCA et al. [256] and MIZUGUCHI et. al [257] first incorporated an enzyme in a fuel cell, which was termed as enzyme fuel cell. Since then many investigations have been reported concerning enzyme fuel cells [258–261].

In the mid 1970s, immobilization of enzymes and microbial cells to solid matrices were enormously developed. The preparation of insoluble heterogeneous

78

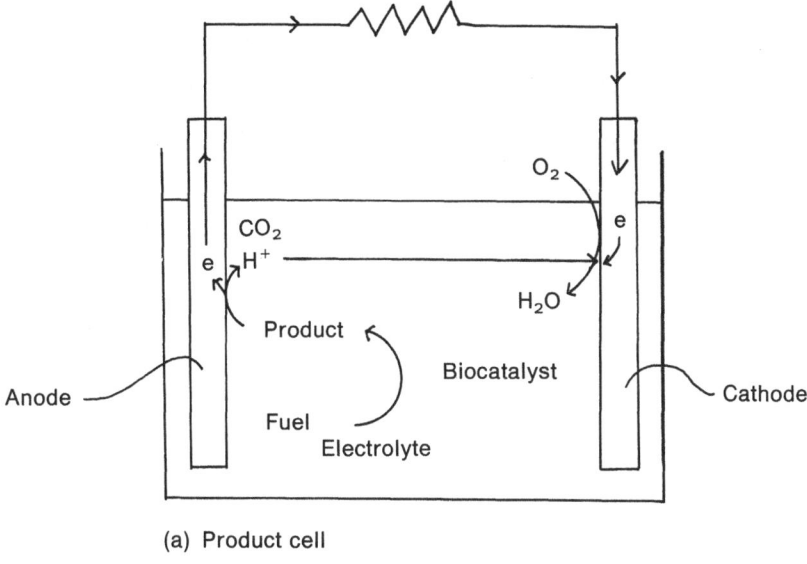

(a) Product cell

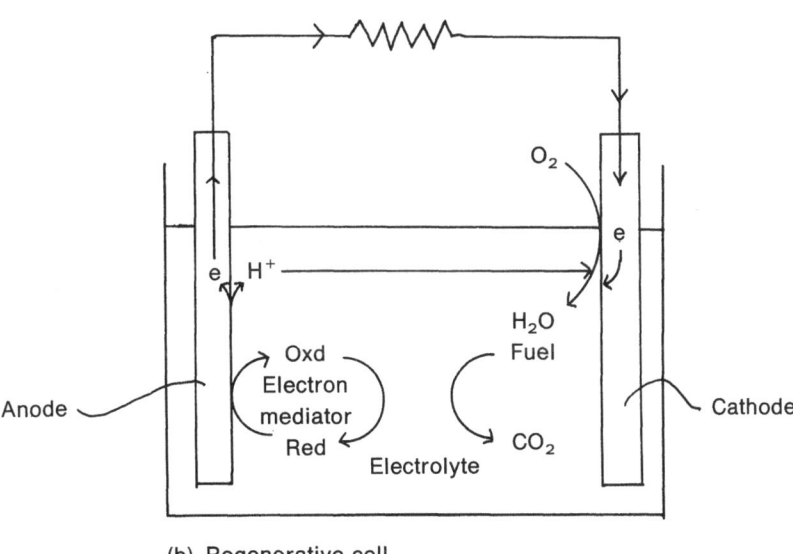

(b) Regenerative cell

Fig. 1.52. Principle of enzyme fuel cells

biocatalysts has many advantages, such as easy separation from the reaction mixture, the possibility of continuous catalysis, and an increase in the stability of enzymes. Immobilized enzymes and microbial cells have been used to construct bioelectrochemical energy conversion systems such as enzyme fuel cells. KARUBE et al. reported a microbial fuel cell with immobilized microbial cells [258], which

1911	Potter	Potentiometric determination of yeast activity
1931	Cohen	Microbial fuel cell
1962	Rohrback	Microbial fuel cell with hydrogen-producing *Cl. butyricum*
1963	DelDuca	Enzyme fuel cell with urease
1964	Mizuguchi	Enzyme fuel cell with hydrogenase
1964	Berk	Microbial solar cell with photosynthetic bacteria
1970	Takahashi	Enzyme fuel cell with NAD
1977	Karube	Microbial fuel cell with immobilized *Cl. butyricum*

Fig. 1.53. Brief history of enzyme fuel cells

opened a new stage of research and investigations of bioelectrochemical energy conversion [262–264].

A brief historical survey is given in Fig. 1.53 [268].

1.5.2.2 Product Cells

Fundamentally, a fuel cell functions by the oxidation of a highly reduced fuel at the anode, with concomitant transfer of electrons and protons to a suitable reaction, e.g., oxygen to water, at the cathode. In the product cells, electroactive substances are produced by biocatalytic reactions. Hydrogen, formic acid, ethanol, and ammonia may be electroactive at the anode. Product cells are listed in Table 1.18.

Table 1.18. Product cells with enzymes and microbial cells

Biocatalyst	Fuel	Anode	Cathode	Remarks	Ref.
Cl. butyricum	Glucose → H_2	Pd	Pt	$4\,mA \cdot cm^{-2}$	Rohrback (1962)
E. coli	*Lactose* → H_2	Pt	Pt	4–$7\,\mu A \cdot cm^{-2}$	Allen (1972)
Urease	Urea → NH_3			28 V, 0.7 A (64 cells)	DelDuca (1963)
B. pasteurii	Urea → NH_3			1–$3\,mA \cdot cm^{-2}$	Brake (1963)
P. methanica	Methane → ?			$3\,\mu A \cdot cm^{-2}$	Hees (1965)
Nocardia	Ethane → ?			–	Davis (1962)
Immobilized *Cl. butyricum*	Glucose → H_2 HCOOH	Pt black		$1.2\,mA \cdot cm^{-2}$	Karube (1977)

Table 1.19. Regenerative type of enzyme fuel cells

Fuel	Biocatalyst	Electroactive substance	Remarks
H_2	Hydrogenase	Methylene blue	Mizuguchi [257] Kimura [262]
G-6-P	G-6-P dehydrogenase	$NADP^+$	Takahashi [265]
Glucose	*E. coli*	$Fe(CN)_6^{3-}/Fe(CN)_6^{4-}$	Higgins [267]
C_2H_5OH	Alcohol dehydrogenase	NAD^+, FMN	Suzuki [266]
C_2H_5OH	*Acetobacter*	$Fe(CN)_6^{3-}/Fe(CN)_6^{4-}$	Higgins [267]

1.5.2.3 Regenerative Cells

In the regenerative fuel cell, disrupted cells or enzymes are used to regenerate redox compounds that in turn, take part in electrochemical reactions. Some of those reported are given in Table 1.19.

References

1. Bard AJ, Faulkner LR (1980) Electrochemical methods. Wiley, New York
2. Kissinger PT, Heineman WR (eds) (1984) Laboratory techniques in electroanalytical chemistry. Dekker, New York
3. Fujishima A, Aizawa M, Inoue T (1984) In: Electrochemical measurements, vols I and II. Gihodo, Tokyo
4. Dryhurst G, Kadish KM, Scheller F, Renneberg R (1982) Biological electrochemistry. Academic Press, New York, p 409
5. Malmstrom BG (1980) In: Spiro TG (ed) Metal ion activation of dioxygen. Wiley, New York, p 181
6. Tanaka T (1975) J Biochem 77:147
7. Tagawa K, Arnon DI (1968) Biochim Biophys Acta 153:602
8. Deisenhofer J, Epp O, Miki K, Huber R, Michel H (1984) J Mol Biol 180:385
9. Mortenson LE (1978) Biochimie 60:219
10. Kaya RC, Stonehill HI (1952) J Chem Soc 3244
11. Cattuthers C, Tech J (1954) Arch Biochem Biophys 56:441
12. Ke B (1956) Biochim Biophys Acta 20:547
13. Rodkey FL (1959) J Biol Chem 234:188
14. Burnett JN, Underwood AL (1965) Biochemistry 4:2060
15. Cunningham AJ, Underwood AL (1966) Arch Biochem Biophys 117:88
16. Wilson AM, Epple PG (1966) Biochemistry 5:3170
17. Cunningham AJ, Underwood AL (1967) Biochemistry 6:266
18. Underwood AL, Burnett JN (1973) In: Bard AJ (ed) Electroanalytical chemistry, vol 6. Dekker, New York, p 1
19. Aizawa M, Ikariyama Y, Suzuki S (1976) J Solid-Phase Biochem 1:249
20. Aizawa M, Ikariyama Y, Suzuki S (1976) J Solid-Phase Biochem 1:197
21. Santhanam KS, Schmakel CO, Elving PJ (1974) Biochem Bioenerg 1:147
22. Schmakel CO, Santhanam KS, Elving PJ (1975) J Am Chem Soc 97:5083
23. Bresnahan WT, Elving PJ (1981) J Am Chem Soc 103:2379
24. Careli V, Liveratore F, Casini A, Mondelli R, Ainone A, Carelli I, Rotilio G, Mavelli I (1980) J Bioorg Chem 9:497
25. Czocharalska B, Szureykowska M, Shuger D (1980) Arch Biochem Biophys 199:497
26. Jaegfeldt H (1981) Bioelectrochem Bioenerg 8:355
27. Bresnahan WT, Elving PJ (1981) Biochim Biophys Acta 678:151
28. Aizawa M, Coughlin RW, Charles M (1976) Biotechnol Bioeng 18:209
29. Aizawa M, Coughlin RW, Charles M (1976) Biochim Biophys Acta 440:233
30. Aizawa M, Suzuki S, Kubo M (1976) Biochim Biophys Acta 444:886
31. Webber A, Shah M, Osteryoung J (1984) Anal Chim Acta 157:1
32. Webber A, Osteryoung J (1984) Anal Chim Acta 157:7
33. DiCosimo R, Wong CH, Daniels L, Whitesides GH (1981) J Org Chem 46:4622
34. Maeda H, Kajiwara K (1987) Biotechnol Bioeng (in press)
35. Leduc P, Thevenot D (1974) Bioelectrochem Bioenerg 1:96
36. Bladele WJ, Hass RG (1970) Anal Chem 42:918
37. Braun RD, Santhanam KSV, Elving PJ (1975) J Am Chem Soc 97:2591
38. Kelly RM, Kirwan DJ (1977) Biotechnol Bioeng 19:1215
39. Aizawa M, Coughlin RW, Charles M (1975) Biochim Biophys Acta 385:362
40. Coughlin RW, Aizawa M, Alexander BF, Charles M (1975) Biotechnol Bioeng 17:515
41. Wallace TC, Coughlin RW (1977) Anal Biochem 80:133

42. Bladel WJ, Jenkins RA (1975) Anal Chem 47:1337
43. Wallace TC, Leh MB, Coughlin RW (1977) Biotechnol Bioeng 19:901
44. Malinawskas A, Kulys YY (1978) Anal Chim Acta 98:31
45. Bladel WJ, Engstrom RC (1980) Anal Chem 52:1681
46. Jaegfeldt H, Kuwana J, Johansson G (1983) J Am Chem Soc 105:1805
47. Huck H, Schumidt HL (1981) Angew Chem Internatl Ed Engl 20:402
48. Tse DCS, Kuwana T (1978) Anal Chem 50:1315
49. Degrand C, Miller LL (1980) J Am Chem Soc 102:5728
50. Jaegfeldt H, Torstensson A, Gorter L, Johansson G (1981) Anal Chem 53:1979
51. Huck H, Schelte-Graf A, Danzer D, Kirch P, Schumidt HL (1984) Analyst 109:149
52. King TE (1963) J Biol Chem 238:4032
53. Albery WJ, Barlett PN (1984) J Chem Soc Chem Commun 234
54. DiCosino R, Wong CH, Daniels L, Whitesides GM (1981) J Org Chem 46:4100
55. Kuznetsov BA, Mestechkina NE (1977) Bioelectrochem Bioenerg 4:1
56. Scheller F, Stand G, Neumann B, Kuhn M, Ostrowski W (1978) Bioelectrochem Bioeng 6:117
57. Kuzunetsov BA, Shumakovich GP, Mestechkina NE (1977) Bioelectrochem Bioeng 4:512
58. Ikeda T, Ando S, Senda M (1981) Bull Chem Soc Jpn 54:2189
59. Ando S, Ikeda T, Senda M (1980) Rev Polarogr Kyoto 26:18
60. Kakutani T, Senda M (1980) Bull Chem Soc Jpn 53:1942
61. Scheller F, Rennenberg R, Stand G, Pommerening K, Mohr P (1977) Bioelectrochem Bioeng 4:500
62. Archakov AI, Kuznetsov BA, Izotov MV, Karuziva II (1981) Dokl Akad Nauk SSSR 258:216
63. Besto SR, Klapper MH, Anderson LB (1972) J Am Chem Soc 94:8197
64. Scheller F, Janchen M, Etzold G, Will H (1974) Bioelectrochem Bioeng 1:478
65. Scheller F, Janchen M, Lampe J, Prumke HJ, Blanck J, Palecek E (1975) Biochim Biophys Acta 412:157
66. Scheller F, Prumke HJ (1976) J Electroanal Chem 70:219
67. Kuznetsov BA, Mestechkina NE (1977) Bioelectrochem Bioeng 4:1
68. Kuznetsov BA (1981) Bioelectrochem Bioeng 8:681
69. Haladjian J, Bianco P, Serre PA (1979) Bioelectrochem Bioeng 6:555; (1980) J Electroanal Chem 106:397
70. Serre PA, Haladjian J, Bianco P (1981) J Electroanal Chem 122:327
71. Kano T, Nakamura S (1958) Bull Agr Chem Soc Jpn 22:399
72. Yeh P, Kuwana T (1975) Chem Lett 1145
73. Eddows MJ, Hill HAO (1977) J Chem Soc Chem Commun 771; (1979) J Am Chem Soc 101:4461
74. Taniguchi I, Murakami T, Toyosawa T, Yamaguchi H, Yasukouchi K (1972) J Electroanal Chem 131:397
75. Taniguchi I, Toyosawa K, Yamaguchi H, Yasukouchi K (1982) J Electroanal Chem 140:187
76. Cotton TM, Schultz SG, Van Duyne RP (1980) J Am Chem Soc 102:7960
77. Niki K, Yagi T, Inokuchi H, Kimura K (1977) J Electrochem Soc 124:1889
78. Niki K, Yagi T, Inokuchi H, Kimura K (1979) J Am Chem Soc 101:3335
79. Niki K, Tokizawa Y, Kumagai H, Fujiware R, Yagi T, Inokuchi H (1981) Biochim Biophys Acta 636:136
80. Higuchi Y, Bando S, Kusunoki M, Matsuura Y, Ysuoka N, Kakudo M, Yamanaka T, Yagi T, Inokuchi H (1981) J Biochem 89:1659
81. Bianco P, Faugue G (1979) Biochim Biophys Acta 545:86
82. Bianco P, Faugue G, Haladjian J (1979) Bioelectrochem Bioeng 6:385
83. Bianco P, Haladjian J (1981) Electrochem Acta 26:1317
84. Bianco P, Haladjian J (1981) Bioelectrochem Bioeng 8:239
85. Weitzman PDJ, Kenedy IR, Caldwell RA (1971) FEBS Lett 17:241
86. Dalton H, Zubieta J (1973) Biochim Biophys Acta 332:133
87. Chien YW (1976) J Pharm Sci 65:1471
88. Bianco P, Haladjian J (1977) Biochem Biophys Res Commun 78:323

89. Feinberg BA, Lau YK (1980) Bioelectrochem Bioeng 7:187
90. Miller IR, Werber MM (1979) J Electroanal Chem 100:103
91. Cecil R, Weitzman PDJ (1964) Biochem J 93:1
92. Eddows MJ, Hill HAO, Uosaki K (1979) J Am Chem Soc 101:7113
93. Aizawa M, Yabuki S, Shinohara H (1985) In: Torii S (ed) Recent Advances in Electroorganic Chemistry. Studies in Organic Chemistry 30, Elsevier, Amsterdam, p 353
94. Moses PR, Wier L, Murray RW (1975) Anal Chem 47:1882
95. Watkins BF, Behling JR, Kariv E (1975) Anal Chem 47:1882
96. Lane RF, Hubbard AT (1973) J Phys Chem 77:1401; (1973) 77:1411; (1975) 79:808
97. Merz A, Bard AJ (1978) J Am Chem Soc 100:3222
98. Zah J, Kuwana T (1983) J Electroanal Chem 150:645
99. Miller JS (ed) (1982) Chemically modified surfaces in catalysis and electroanalysis. ACS Symposium Series 192, ACS, Washington
100. Fujihira M (1986) In: Fry AJ, Britton E (eds) Topics in organic electrochemistry. Plenum Press, New York, p 255
101. Kobayashi N, Matsue T, Fujihira M, Osa T (1979) J Electroanal Chem 103:427
102. Colman JP, Marrocco M, Cenisevich P, Koval C, Anson FC (1979) J Electroanal Chem 101:117
103. Rocklin RD, Murray RW (1981) J Phys Chem 85:2104
104. Lewis NS, Wrighton MS (1981) Science 211:944
105. Landrum HL, Salmon RT, Hawkridge FM (1977) J Am Chem Soc 99:3154
106. Bowden EF, Hawkridge FM, Blount HN (1980) Bioelectrochem Bioeng 7:477
107. Crawley CD, Hawkridge FM (1981) Biochem Biophys Res Commun 99:516
108. Albery WJ, Eddowes MJ, Hill HAO, Hillman AR (1981) J Am Chem Soc 103:3904
109. Hill HAO, Walton NJ, Higgins IJ (1981) FEBS Lett 126:282
110. Higgins IJ, Hill HAO (1978) UK Patent Application No. 2 033 428
111. Kulys JJ, Cenas NK, Svirmickas GJS, Svirmickiene VP (1982) Anal Chim Acta 138:19
112. Ianniello RM, Lindsay TJ, Yacynych AM (1982) Anal Chem 54:1098
113. Ianniello RM, Lindsay TJ, Yacynych AM (1982) Anal Chem Acta 141:23
114. Shu FR, Wilson GS (1976) Anal Chem 48:1679
115. Kamin RA, Wilson GS (1980) Anal Chem 52:1198
116. Bowdillon C, Bourgeois JP, Thomas D (1980) J Am Chem Soc 102:423
117. Ianniello RM, Yacynych AM (1981) Anal Chem 53:2090
118. Katasho I, Ikeda T, Senda M (1982) Rev Polarogr Kyoto 28:74
119. Aizawa M, Coughlin RW, Charles M (1975) Biochim Biophys Acta 385:362
120. Coughlin RW, Aizawa M, Charles M, Alexander BR (1975) Biotechnol Bioeng 17:515
121. Burnett JN, Underwood AL (1973) In: Electroanalytical chemistry, vol 6. Dekker, New York
122. Janik B, Elving PJ (1968) Chem Rev 68:295
123. Aizawa M, Suzuki S, Kubo M (1976) Biochim Biophys Acta 444:886
124. Aizawa M, Namba K, Suzuki S (1977) Arch Biochem Biophys 180:41
125. Aizawa M, Namba K, Suzuki S (1977) Arch Biochem Biophys 182:305
126. Karube I, Nakamoto Y, Namba K, Suzuki S (1900) Biochim Biophys Acta 429:975
127. Karube I, Nakamoto Y, Suzuki S (1976) Biochim Biophys Acta 445:774
128. Karube I, Yugeta Y, Suzuki S (1977) Biotechnol Bioeng 19:1493
129. Matsuoka H, Suzuki S, Aizawa M, Kimura Y, Ikegami A (1981) J Appl Biochem 3:437
130. Shinohara H, Aizawa M, Shirakawa H (1986) J Chem Soc Chem Commun 87
131. Shinohara H, Aizawa M, Shirakawa H (1985) Chem Lett 179
132. Seegopaul P, Rechnitz GA (1983) Anal Chem 55:1929
133. Seegopaul P, Rechnitz GA (1983) Anal Chim Acta 151:91
134. Seegopaul P, Rechnitz GA (1982) Anal Lett 15:709
135. Hassan SS, Rechnitz GA (1982) Anal Chem 54:303
136. Plese C, Fox W, Wiliams K (1983) Clin Chem 29:407
137. Mascini M, Palleschi G, D'ottavio D, Mazzella G (1983) Ann Chim 73:29
138. Gnanasekaran R, Mottola HA (1985) Anal Chem 57:1005
139. Mottola HA (1983) Anal Chim Acta 145:27
140. Rochs R, Riley C (1982) Clin Chem 28:409

141. Elving PJ, Schmakel CO, Santhanam KSV (1976) Crit Rev Anal Chem 6:1
142. Elving PJ, Bresnaha WT, Moiroux J, Semec Z (1982) Bioelectrochem Bioeng 9:365
143. Kitami A, So Y, Miller LL (1981) J Am Chem Soc 103:7636
144. Thoman LC, Christian GD (1975) Anal Chim Acta 78:271
145. Guilbault GG, Cserfalvi T (1975) Anal Chim Acta 78:271
146. Moiroux J, Elving PJ (1976) Anal Chem 51:346
147. Christian GD (1984) In: Ion and enzyme electrodes in biology and medicine. International Workshop, University Park Press, Baltimore, pp 173–181
148. Shah M, Osteryoung J (1982) Anal Chem 54:586
149. David GC, Holland KL, Kissinger PT (1979) J Liquid Chromatogr 2:663
150. Laval JM, Bourdillon C, Moiroux J (1984) J Am Chem Soc 106:4701
151. Jaegfeldt H, Tarstensson A, Johnsson G (1978) Anal Chim Acta 97:221
152. Gladyshev PP, Galdyshev VP, Mamleev VS, Kovalyova SV (1981) Mikrochim Acta 2:289
153. Yao T, Musha S (1979) Anal Chim Acta 110:203
154. Weber SG, Purdy WC (1979) Anal Lett 12:1
155. Gleria KD, Hill HAO, McNeil CJ (1986) Anal Chem 58:1203
156. Heineman WR, Anderson CW, Halsall HB (1979) Science 204:865
157. Wehmeyer KR, Halsall HB, Heineman WR (1982) Clin Chem 28:1968
158. Alam IA, Christian GD (1982) Anal Lett 15(B18):1449
159. Doyle MJ, Halsall HB, Heinemann WR (1982) Anal Chem 54:2318
160. Rigo A, Viglino P (1975) Anal Biochem 68:1
161. Cheny FS, Christian GD (1977) Anal Chem 49:1785
162. Johnson JM, Halsall HB, Heineman WR (1982) Anal Chem 54:1377
163. Aizawa M, Morioka A, Matsuoka H, Suzuki S, Nagamura Y, Shinohara R, Ishiguro I (1976) J Solid-Phase Biochem 1:319
164. Aizawa M, Morioka A, Suzuki S (1980) Anal Chim Acta 115:61
165. Yuan C, Kuan SS, Guilbault GG (1981) Anal Chem 53:190
166. Alexander PW, Maltra C (1982) Anal Chem 54:68
167. Meyerhoff ME, Rechnitz GA (1979) Anal Biochem 95:483
168. Clark LC Jr, Lyons C (1962) Ann NY Acad Sci 102:29
169. Updike SJ, Hicks GP (1967) Nature (London) 214:986
170. Guilbault GG (1976) Handbook of enzymatic methods of analysis. Dekker, New York
171. Suzuki S (ed) (1983) Biosensors. Kodansha, Tokyo
172. Aizawa M (1983) In: Seiyama T, Suzuki S, Shiokawa S, Fueki K (eds) Chemical sensors. Kodansha-Elsevier, p 683
173. Scheller FW, Schunbert F, Rennenberg R, Muller HG (1985) Biosensors 1:135
174. Koyama M, Sato Y, Aizawa M, Suzuki S (1980) Anal Chim Acta 116:307
175. Aizawa M (1984) In: Lloyd DR (ed) Materials science of synthetic membranes. Am Chem Soc, Washington DC, p 447
176. Shinohara H, Chiba T, Aizawa M (1987) Sensors and actuators (in press)
177. Cass AEG, Davis G, Francis GD, Hill HAO, Aston WJ, Higgins IJ, Plotkin EV, Scott LDL, Turner APF (1984) Anal Chem 56:1880
178. Shichiri M, Kawamori R, Goriya Y, Yamasaki Y, Nomura M, Hakui N, Abe H (1983) Diabetologia 24:179
179. Gronow M, Turton M (1984) In: Mark V (ed) Clinical biochemistry nearer the patient. Churchill, Livingstone, p 108
180. Cleland N, Enfors SO (1984) Anal Chem 56:1880
181. Rennenberg R, Scheller F (1983) In: Seiyama T, Suzuki S, Shiokawa J, Fueki K (eds) Chemical sensors. Kodansha-Elsevier, p 711
182. Karube I, Matsuoka H, Suzuki S, Watanabe E, Toyama K (1984) J Argic Food Chem 32:314
183. Kubo I, Osawa H, Karube I, Matsuoka H, Suzuki S (1983) In: Seiyama T, Suzuki S, Shiokawa J, Fueki K (eds) Chemical sensors. Kodansha-Elsevier, p 660
184. Thevenot DR, Sternberg R, Coulet PR, Laurent J, Gautheron G (1979) Anal Chem 51:96
185. Yao T (1983) Anal Chim Acta 148:27
186. Mizutani F, Shimura Y, Tsuda K (1984) Chem Lett 199

187. Booker H, Haslam J (1974) Anal Chem 46:1054
188. Mizutani F, Sasaki K, Shimura Y (1983) Anal Chem 55:35
189. Mizutani F, Sasaki K, Shimura Y, Tsuda K (1983) Seiyama T, Suzuki S, Shiokawa J, Fueki K (eds) Chemical sensors. Kodansha-Elsevier, p 644
190. Rennenberg R, Riedel K, Liebs P, Scheller F (1984) Anal Lett 17(B35):349
191. Niwa M (1983) Clin Chem 29:1177
192. Mosbach K, Danielson B (1974) Biochim Biophys Acta 364:140
193. Mosbach K, Danielson B (1981) Anal Chem 53:83A
194. Bergveld P (1970) IEEE Trans BME-17(1):70
195. Matsuo T, Wise KD (1974) IEEE Trans BME-21(6):485
196. Moss SD, Janata J, Johnson CC (1975) Anal Chem 47:2238
197. Caras S, Janata J (1980) Anal Chem 52:1935
198. Miyahara Y, Moriizumi T, Shiokawa S, Matsuoka H, Karube I, Suzuki S (1983) J Chem Soc Jpn 823
199. Kuriyama T, Kimura J, Kawana Y (1984) 16th International Conference of Solid State Devices and Materials. Kobe, LC-12-3
200. Lundstorm I (1981) Sensors and Actuators 1:403
201. Winquist F, Spetz A, Lundstrom I, Danielsson B (1984) Anal Chim Acta 163:143
202. Aizawa M, Ikariyama Y, Kuno H (1984) Anal Lett 17(B7):555
203. Seitz WR (1984) Anal Chem 56:16A
204. Saari LA, Seitz WR (1982) Anal Chem 54:821
205. Peterson JI, Fitzgerald RV, Buckhold DK (1984) Anal Chem 56:62
206. Ikariyama Y, Aizawa M, Suzuki S (1980) J Solid-Phase Biochem 5:223
207. Aizawa M, Suzuki S, Kato T, Fujiwara Y, Fujita Y (1980) J Appl Biochem 2:190
208. Ikariyama Y, Suzuki S, Aizawa M (1984) Anal Chim Acta 156:245
209. Tien HT (1974) In: Bilayer lipid membrane (BLM): theory and practice. Dekker, New York
210. Bangham AD, Hill MW, Miller NGA (1974) In: Korn ED (ed) Methods in membrane biology, vol 1. Plenum Press, New York
211. Tien HT (1968) Nature 219:272; (1968) J Phys Chem 72:4512
212. Steinemann A, Stark G, Laeuger P (1972) J Memb Chem 9:177; Cherry RJ, Hsu K, Chapmen D (1972) Biochim Biophys Acta 288:12; Weller H, Tien HT Biochim Biophys Acta
213. Tributsch H, Calvin M (1971) Photochem Photobiol 14:95
214. Takahashi F, Kikuchi R (1976) Biochim Biophys Acta 430:490
215. Takahashi F, Aizawa M, Kikuchi R, Suzuki S (1977) Electrochim Acta 22:289
216. Fong FK, Winograd N (1976) J Am Chem Soc 98:2287
217. Fuge DR, Fong GD, Fong FK (1979) J Am Chem Soc 101:3694
218. Aizawa M, Hirano M, Suzuki S (1978) Electrochim Acta 23:1185
219. Aizawa M, Hirano M, Suzuki S (1978) J Memb Sci 4:251
220. Aizawa M, Hirano M, Suzuki S (1979) Electrochim Acta 24:89
221. Aizawa M, Yoshitake J, Suzuki S (1981) Mol Cryst Liq Cryst 70:129
222. Miyasaka T, Watanabe T, Fujishima A, Honda K (1978) J Am Chem Soc 1000:6647
223. Aizawa M, Shinohara H, Watanabe S (1982) Denki Kagaku 50:854
224. Dancshazy Z, Karvaly B (1976) FEBS Lett 72:136
225. Herrmann TR, Rayfield GW (1976) Biochim Biophys Res Commun 71:603
226. Shieh P, Packer L (1976) Biochem Biophys Res Commun 71:603
227. Skulachev VP (1977) FEBS Lett 64:23
228. Packer L, Konishi T, Shieh P (1977) In: Buvet R et al. (eds) Living systems as energy converters. Elsevier, Amsterdam, p 119
229. Tang CW, Albercht AC (1974) Mol Cryst Liq Cryst 25:53
230. Langmuir I, Schafer VJ (1937) J Am Chem Soc 59:2075
231. Hanson EA (1939) Rec Trv Botan Beerl 36:180
232. Bellamy WD, Gaines GL, Tweet AG (1963) J Chem Phys 39:2528
233. Shieh P, Tien HT (1974) J Bioenerg 6:45
234. Tien HT (1979) In: Baker J (ed) Photosynthesis in relation to model systems. Elsevier, Amsterdam, p 116

235. Hong FT (1976) Photochem Photobiol 24:155
236. Gross EL, Youngman DR, Winemiller SL (1978) Photochem Photobiol 28:249
237. Ochiai H, Shibata H, Matsuo T, Hashinokuchi K, Inamura I (1978) Agric Biol Chem 42:683
238. Miyasaka T, Honda K (1980) In: ACS Sympsoium series: Photoeffects at semiconductor-electrolyte interfaces. American Chemical Society, Washington DC, p 1
239. Shiozawa H, Kobayashi H, Kurihara K, Iida T, Mitamura T (1981) J Chem Soc Jpn 1057
240. Iida T, Shiozawa H, Kobayashi H, Mitamura T (1982) Agric Biol Chem 46:275
241. Ochiai H, Shibata H, Fujishima A, Honda K (1979) Agric Biol Chem 43:881
242. Ochiai H, Shibata H, Sawa Y (1980) Proc Natl Acad Sci USA 77:2442
243. Arnon DI, Mitsui A, Paneque A (1961) Science 134:1425
244. Benemann JR, Berenson JA, Kaplan NO, Kamen MD (1973) Proc Natl Acad Sci USA 70:2317
245. Yagi T (1976) Proc Natl Acad Sci USA 73:2947
246. Yagi T, Goto M, Nakano K, Kimura K, Inokuchi H (1975) J Biochem 78:443
247. Karube I, Aizawa K, Ikebe S, Suzuki S (1979) Biotechnol Bioeng 21:253
248. Karube I, Otsuka T, Kayano H, Matsunaga T, Suzuki S (1980) Biotechnol Bioeng 22:2655
249. Suzuki S, Karube I, Matsunaga T (1979) Biotechnol Bioeng Symp 8:501
250. Kayano H, Matsunaga T, Karube I, Suzuki S (1981) Biotechnol Bioeng 23:2283
251. Kayano H, Karube I, Matsunaga T, Suzuki S, Nakayama O (1981) Eur J Appl Microbial Biotechnol 121:1
252. Hill HAO, Higgins IJ (1981) Phil Trans Royal Soc London A302
253. Potter MC (1912) Proc Roy Soc (London) Ser B 84:260
254. Cohen B (1931) J Bacteriol 21:18
255. Rohrback GH, Scott WR, Canfield JH (1962) Proceeding of 16th Annual Power Source Conference, p 18
256. DelDuca DG, Fuscoe JM, Zurilla RW (1963) Develop Ind Microbiol 4:81
257. Mizuguchi J, Suzuki S, Kashiwaya K, Tokura M (1964) Kogyo Kagaku Zasshi 67:410
258. Karube I, Matsunaga T, Tsuru S, Suzuki S (1977) Biotechnol Bioeng 19:1727
259. Brake J, Momyer W (1963) 3rd Quarterly Progress Report 4:1
260. van Hess W (1965) J Electrochem Soc 112:158
261. Davis JB, Yarbrough HF Jr (1962) Science 137:615
262. Kimura K, Inokuchi H, Yagi T (1972) Chem Lett 693
263. Canfield JH, Coldner BH (1964) Final report. Magna Corp., Contact No. NASw-623 1964
264. Videla HA, Arvia AJ (1975) Biotechnol Bioeng 17:1529
265. Takahashi F, Aizawa M, Mizuguchi J, Suzuki S (1970) Kogyo Kagaku Zasshi 73:62
266. Suzuki S, Takahashi F, Saitoh I, Sonobe N (1975) Bull Chem Soc Japan 48:3246
267. Higgins IJ, Hammond RG, Plotkin E, Hill HAO, Uosaki K, Eddowes MJ, Cass AEG (1980) In: Harrison DEF et al. (eds) Hydrocarbons in biotechnology. Heydon & Son, London, pp 181–194
268. Allen MJ (1972) In: Method in microbiology. Academic Press, New York, p 247

2 Coenzyme Regeneration

2.1 Introduction

In the last two decades, development of the immobilized enzyme technique has enabled repeated and continuous use of expensive water-soluble enzymes with improved stability and now this technique is widely used in the various fields of biochemical conversion, optical resolution, and analytical use. However, practical applications of this technique on an industrial scale is still restricted to hydrolases, lyases, and isomerases as shown by the following examples [2, 11]:

(1) Resolution of racemic mixtures of amino acids by immobilized amino acylase
(2) L-Aspartic acid by immobilized cells containing aspartase
(3) 6-Aminopenicillanic acid by immobilized penicillin amidase
(4) Fructose by immobilized glucose isomerase
(5) L-Malic acid by immobilized cells containing fumarase
(6) Nonlactose milk by immobilized lactase
(7) D-Amino acid derivatives by immobilized hydantoinase

Although other types of enzymes such as oxidoreductases, transferases, and ligases have strong potentials for synthetic purposes in making various chemicals and pharmaceuticals and for analytical use, enzymes in these groups are still unused in bioreactors mainly because of their needs for cofactors.

The cofactors are classified into three groups: (1) inorganic cofactors (metal), (2) prosthetic groups, and (3) dissociable coenzymes. Inorganic cofactors and prosthetic groups are strongly bound to enzyme proteins so that the immobilization of enzyme proteins simultaneously corresponds to the stable immobilization of these cofactors. Dissociable coenzymes, however, are loosely bound to the apoenzyme and stoichiometrically consumed like substrates in the enzyme reactions. Coenzymes are generally very expensive, so their stoichiometric consumption is economically not practical. It is well known that over one third of the two thousand enzymes registered by the International Union of Biochemistry require one of the five adenine coenzymes (NAD, NADP, ATP, FAD, and CoA) for the expression of their catalytic activity [7]. These facts explain the increasing interests in the development of methods for efficient coenzyme regeneration and recycling [1–10].

This chapter is devoted to the three major dissociable coenzymes: nicotinamide adenine dinucleotide (NAD), nicotinamide adenine dinucleotide phosphate

Table 2.1. Structure and function of NAD, NADP, and ATP

Coenzyme	Structure		Molecular weight	Entity transferred
	R_1	R_2		
NAD	A	OH	663	Hydrogen (electron)
NADP	A	P	743	Hydrogen (electron)
ATP	P	OH	507	Phosphate with energy

Residue A

(NADP), and adenosine 5′-phosphate (ATP). Structures and functions of these coenzymes are listed in Table 2.1. Pyridine nucleotides, NAD and NADP, are electron acceptors and the coenzymes of a large number of oxidoreductases, collectively called pyridine-linked dehydrogenases. ATP primarily is a carrier of phosphate and pyrophosphate in several important enzymatic reactions involved in the transfer of chemical energy. Other nucleoside triphosphates (NTP), GTP, CTP, and UTP, also act as alternatives for chemical energy supply into specific biosynthetic pathways [61]. Fortunately, some enzymes for the regeneration of ATP also accept other NTPs as coenzymes [9]. Therefore, the regeneration problems of all NTPs may be treated in the same manner.

Other dissociable coenzymes such as coenzyme A (CoA), S-adenosylmethionine (SAM) and 3′-phosphoadenosine-5′-phosphosulfate (PAPS) do not seem to be very important in industrial use in spite of their importance in the metabolic pathways [9].

2.2 Methods for Coenzyme Regeneration

2.2.1 Reduced Form of Pyridine Nucleotides (NADH, NADPH)

The least expensive nicotinamide coenzyme, NAD, costs about 1000 US$/mol [9]. The other nicotinamide coenzymes, NADH, NADP, and NADPH, presumably are about twice, six, and twenty times as expensive as NAD, respectively. These figures clearly show the potential needs for regeneration of these coenzymes for repeated use.

The energy potential of the reduced state of pyridine nucleotides is higher than that of the oxidized state. This makes the regeneration from the oxidized to the reduced form more difficult than in the opposite case from the reduced to the oxidized state. Various methods are available for this purpose: (1) chemical, (2) electrochemical, (3) enzymatic, and (4) photobiochemical methods.

As for the chemical method, JONES et al. [62] used sodium dithionite as a reducing agent of NAD^+ to NADH. As sodium dithionite is a strong reducing agent, NAD^+ is favorably reduced to NADH thermodynamically. The problem with this method, however, exists in the instability of this compound in aqueous solutions and the possible inhibitory and inactivating effects of this compound towards enzymes [23].

TAYLOR and JONES [68] applied the reaction between the hydropyridine agents and NAD^+. Various functional groups were introduced to change the oxidation-reduction potential of the hydropyridines. Hydropyridines with a standard redox potential more negative than that of NAD^+/NADH (-315 mV) were available for the regeneration of NADH.

The electrochemical method has an advantage of not requiring specific substrates. Its disadvantages, however, are the existence of high overpotentials in the electrochemical reaction and the occurrence of a nonspecific reaction; the inactive dimer of NAD is formed in the electrode reaction. Because of this, the yield of enzymatically active NADH from NAD^+ is low in the direct electrochemical reduction. To avoid dimerization, attempts with a macromolecule-bound NAD or a liquid-crystal electrode were carried out [18–21]. To reduce the high overpotential, a mediator such as methyl viologen is frequently used. Enzymes such as diaphorase or lipoamide dehydrogenase are also used to improve the specificity and the kinetics of the transhydrogenation reaction between the mediator and the coenzyme. More details on these topics were already given in Chap. 1.

Enzymatic methods of NAD(P)H regeneration are most frequently used because of the excellent specificity and the rapid kinetics. These are listed in Table 2.2. There are four criteria for evaluation of the enzymatic method: (1) equilibrium, (2) specific activity, (3) cost of enzyme, and (4) cost of substrate for the coenzyme regeneration. These criteria also hold in all other cases with $NAD(P)^+$ and ATP.

Alcohol dehydrogenase (ADH) with ethanol is most conveniently used [24, 27, 29, 32] because of the low commercial price of both enzyme and substrate for regeneration. Yeast ADH is one of the least expensive dehydrogenases available (0.6 US\$/1000 U) with a good specific activity (350 U/mg). However, the thermodynamic equilibrium is not favorable for the reduction of $NAD(P)^+$. The equilibrium constant, K_{eq}, in Table 2.2, is defined as:

$$K_{eq} = [product][NADH][H^+]/[substrate][NAD^+]. \qquad (2.1)$$

As this constant includes the effect of pH, the reactions with K_{eq} higher than 10^{-7}–10^{-8} M are favorable for the purpose of NAD(P)H regeneration at neutral pH. The equilibrium constant for ADH reaction corresponds to the driving force $\Delta G = +23.8$ KJ/mol [23]. The overall equilibrium of the ADH reaction may be shifted to a more favorable state by removing the product (aldehyde) through gas-

Table 2.2. Enzymatic regeneration of NAD(P)H

Enzyme (EC)	Origin	Coenzyme	Substrate	Specific activity [U/mg]	Price[a] [$/1000 U]	Equilibrium constant, K_{eq} [M]	Ref.
AlcoholDH (1.1.1.1)	Horse liver	NADH	Ethanol	2.2	130	8.0E−12	[25, 26]
AlcoholDH (1.1.1.1)	Yeast	NADH	Ethanol	352	0.6	8.0E−12	[26, 28]
AlcoholDH (1.1.1.2)	Thermoanaerobium brokii	NADPH	Ethanol	5–15	230	8.0E−12	[31]
AldehydeDH (1.2.1.5)	Bakers yeast	NADH	Acetaldehyde	5–10	260	Irreversible	[71]
AldehydeDH (1.2.1.5)	Saccharomyces cerevisiae	NADH	Propionaldehyde	20.6	–	–	[34]
AldehydeDH (1.2.1.5)	Saccharomyces cerevisiae	NADPH	Propionaldehyde	8.7	–	–	[34]
AldehydeDH (1.2.1.4)	Proteus vulgaris	NADPH	Isovaleraldehyde	14.6	–	–	[33]
GlucoseDH (1.1.1.47)	Bacillus subtilis	NAD(P)H	Glucose	368	300	3E−6	[35]
GalactoseDH (1.1.1.48)	Pseudomonas fluorescens	NADH	Galactose	3–6	2200	1.14E−4	[10]
G6PDH (1.1.1.49)	Saccharomyces carlsbergensis	NADPH	G6P	676	–	6E−7	[37]
G6PDH (1.1.1.49)	Saccharomyces cerevisiae	NADPH	G6P	400	34	6E−7	[38]
G6PDH (1.1.1.49)	Saccharomyces cerevisiae	NADPH	G6S	60	34	–	[38]
G6PDH (1.1.1.49)	Leuconostoc mesenteroides	NADH	G6P	700	29	6E−7	[38, 39]
G6PDH (1.1.1.49)	Leuconostoc mesenteroides	NADPH	G6P	400	29	6E−7	[38, 39]
G6PDH (1.1.1.49)	Leuconostoc mesenteroides	NADH	G6S	2	29	–	[38]
G6PDH (1.1.1.49)	Leuconostoc mesenteroides	NADPH	G6S	10	29	–	[38]
G6PDH (1.1.1.49)	Bacillus stearothermophilus	NADH	G6P	115	220	6E−7	[36]
G6PDH (1.1.1.49)	Bacillus stearothermophilus	NADPH	G6P	100	220	6E−7	[36]
FormateDH (1.2.1.2)	Candida boidini	NADH	Formic acid	2.4	–	Favourable	[41]
FormateDH (1.2.1.2)	Pseudomonas oxalaticus	NADH	Formic acid	0.4–1.2	680	Favourable	[71]
FormateDH (1.2.1.2)	Arthrobacter sp. KM62	NADH	Formic acid	–	–	Favourable	[40]
Hydrogenase (1.18.3.1)	Alcaligenes eutrophus	NADH	H_2	54.5	–	Favourable	[44]
Ferredoxin–NAD reductase (1.18.1.1)	–	NADH	Ferredoxin	–	–	–	[49]

Ferredoxin–NADP reductase (1.18.1.2)	Spinach	NADPH	Ferredoxin	1.5	8500	–	[49]
Malic enzyme (1.1.1.40)	Chicken liver	NADPH	Malic acid	10–30	730	5.1E−2	[71]
AldohexoseDH	Gluconobacter cerinus	NADPH	Glucose etc.	10.1	–	Favourable	[56]
Methylviologen-dependent NAD(P) reductase	Clostridium kluyveri	NADH	MV^+	16.5	–	–	[183]
Methylviologen-dependent NAD(P) reductase	Clostridium kluyveri	NADPH	MV^+	3.3	–	–	[183]

[a] Reagent price in US$.

bubbling [23] or organic solvent extraction [2]. NADP-linked ADH (EC 1.1.1.2) was also isolated [22, 31] and is commercially available though expensive.

The aldehyde dehydrogenase (ALDDH) reaction has a good equilibrium with a large driving force ($\Delta G = -50.2$ KJ/mol) [23], but the commercial price of this enzyme is quite high, while the specific activity is relatively low (5–10 U/mg). This enzyme may be combined with ADH to improve the overall equilibrium [23, 24].

Glucose dehydrogenase (GDH) seems quite attractive with its high specific activity (368 U/mg), the low cost of the substrate, the high driving force ($\Delta G = -23.0$ KJ/mol) [23], and the applicability for both NADH and NADPH. The commercial price of this enzyme, however, is very high. AVIGAD [56] purified aldohexose dehydrogenase from *Gluconobacter cerinus*. This enzyme is NADP-linked with a wide specificity to many aldohexoses such as D-glucose, D-mannose etc.

Galactose dehydrogenase [10] is provided with a good equilibrium but its specific activity is low (3–6 U/mg). The reagent price of this enzyme is very high. The specific activity was recently improved up to 50 U/mg by using recombinant DNA techniques [71].

Glucose-6-phosphate dehydrogenase (G6PDH) has many advantages, such as high specific activity (400–700 U/mg), low commercial price, and applicability both for NAD and NADP. Various types of G6PDH are commercially available, including those from thermophilic bacteria [36]. The only disadvantage with this enzyme is a high cost of the substrate, glucose-6-phosphate (G6P). WONG et al. [38] used glucose-6-sulfate (G6S) in place of G6P. G6S functioned as a substrate with a lower activity than G6P. The stability of the produced NAD(P)H, however, was far better with G6S than with G6P because the former was inactive as an acid catalyst for the decomposition of NAD(P)H. In addition, G6S was more easily prepared than G6P. They concluded that G6S is a preferable reducing agent for regeneration of NADPH.

Formate dehydrogenase (FDH) seems quite useful with the low cost of the substrate (formate), the inertness of the product (CO_2), and the favorable thermodynamic equilibrium ($\Delta G = -19.2$ KJ/mol). However, the specific activity is not satisfactory (0.4–2.4 U/mg) and the commercial price of this enzyme is still high. FDH is linked only to NAD.

WONG et al. [24] used ADH together with ALDDH, FDH. Methanol was the only substrate in the sequence reactions for the regeneration of NADH. In this system, three molecules of NADH were regenerated with consumption of one molecule of methanol. IZUMI et al. [40] used freeze-thawed, air-dried, or acetone-dried cells of *Arthrobacter* containing FDH to reduce NAD^+. EGUCHI et al. [58, 59] used resting cells of methanogen to reduce $NADP^+$. The latter also contained hydrogenase activity.

Hydrogenase, in principle, is quite attractive because the substrate (H_2) and the product (H^+) are inexpensive and inert and the driving force of the reaction is high ($\Delta G = -19.3$ KJ/mol) [23]. However, this enzyme is quite instable and is not commercially available yet. SCHNEIDER et al. [44] studied *Alcaligenes* hydrogenase, which could reduce NAD with consumption of molecular hydrogen at a substantial activity. But this enzyme was instable in its active state under reducing conditions. WONG et al. [49] used *Methanobacterium* hydrogenase [45], which had a good stability but should be combined with ferredoxin reductase or lipoamide dehydrogenase for the reduction of $NAD(P)^+$. Resting cells of *Methanogen* containing both hydrogenase and F_{420}-NADP oxidoreductase were also used for the purpose of reducing $NADP^+$ [59].

Diaphorase (DIA) [50–53], lipoamide dehydrogenase [55, 71], and ferredoxin reductase [49] are frequently used in the mediated electrochemical regeneration of NAD(P)H. These enzymes catalyze the transhydrogenation between a mediator and a coenzyme. As a mediator, methyl viologen, dichlorophenol indophenol, and $K_3Fe(CN)_6$ are frequently used. Diaphorase and lipoamide dehydrogenase with a moderate activity are commercially available at a moderate price (Table 2.3). Ferredoxin reductase is also commercially available but is quite expensive.

Lactate dehydrogenase (LDH), listed in Table 2.3, is not appropriate for the reductive regeneration of pyridine coenzyme because of its unfavorable thermodynamic equilibrium. However, the product of the reductive regeneration reaction (pyruvate) could be chemically reduced back to the original state (lactate) by molecular hydrogen (Fig. 2.5) [16]. This reaction was catalyzed by the bis(phosphine)rhodium complex. Drawbacks of this unique approach were the low activity and the instability of the rhodium catalyst.

A sonicate of *Achromobacter* containing malic enzyme was used for the reduction of $NADP^+$ [60]. The equilibrium of this reaction is quite favorable, but the substrate for the regeneration (malate) is expensive.

As for the photobiochemical method of NADPH regeneration, an immobilized spinach chloroplast was used together with ferredoxin reductase and ferredoxin [57]. $NADP^+$ was continuously reduced to NADPH under illumination for 2 h. This challenging approach still needs improvement of the stability of the enzyme system, though the stability in storage was improved by immobilization.

To conclude, ADH, GDH, G6PDH, and FDH seem to fulfill most criteria required for the regeneration purpose, although each enzyme has each own drawback. DIA and lipoamide dehydrogenase are useful when combined with the electrochemical process. Other enzymatic methods may be of limited interest for practical purposes.

2.2.2 Oxidized Form of Pyridine Nucleotides (NAD⁺, NADP⁺)

Being a down hill reaction, the regeneration of the oxidized form of pyridine nucleotide coenzyme from its reduced state is easier than the reversed case. Chemical, electrochemical, and enzymatic methods are available for this purpose.

In the chemical method, the oxidation-reduction potential of the oxidizing agent of NAD(P)H should be more positive than that of NAD(P)H. Methylene blue and 2,6-dichlorophenol indophenol could oxidize NAD(P)H on irradiation with light [65]. Flavin adenine dinucleotide (FAD) accelerated the reoxidation of the reduced methylene blue by atmospheric oxygen. JONES et al. used pyridinium salt and flavine mononucleotide (FMN) to regenerate NAD^+ from NADH [63, 67–69]. The rate of hydrogen transfer from NADH to the pyridinium salt was positively related to the magnitude of the donor-acceptor redox potential difference [68]. FMN was practically advantageous for its high solubility, low cost, and

Table 2.3. Enzymatic regeneration of NAD(P)[+]

Enzyme (EC)	Origin	Coenzyme	Substrate	Specific activity [U/mg]	Price[a] [$/1000 U]	Equilibrium constant, K_{eq} [M^{-1}]	Ref.
AlcoholDH (1.1.1.1)	Horse liver	NAD[+]	Acetaldehyde	5.9	130	1.3E+11	[26, 25]
AlcoholDH (1.1.1.1)	Yeast	NAD[+]	Acetaldehyde	3250	0.6	1.3E+11	[26, 28]
GlutamateDH (1.4.1.3)	*Tetrahymena pyriformis*	NAD[+]	α-Ketoglutarate, NH$_4^+$	120	–	2.2E+13	[82]
GlutamateDH (1.4.1.3)	*Tetrahymena pyriformis*	NADP[+]	α-Ketoglutarate, NH$_4^+$	120	–	2.2E+13	[82]
GlutamateDH (1.4.1.3)	Bovine liver	NAD[+]	α-Ketoglutarate, NH$_4^+$	60	5	2.2E+13	[10]
GlutamateDH (1.4.1.3)	Bovine liver	NADP[+]	α-Ketoglutarate, NH$_4^+$	–	5	2.2E+13	[10]
GlutamateDH (1.4.1.3)	Frog liver	NAD[+]	α-Ketoglutarate, NH$_4^+$	24	–	2.2E+13	[10]
GlutamateDH (1.4.1.3)	Frog liver	NADP[+]	α-Ketoglutarate, NH$_4^+$	–	–	2.2E+13	[10]
GlutamateDH (1.4.1.4)	*Proteus species*	NADP[+]	α-Ketoglutarate, NH$_4^+$	300	20	2.2E+13	[71]
L-LactateDH (1.1.1.27)	Bovine heart	NAD[+]	Pyruvate	400	3.8	3.6E+11	[10, 83]
L-LactateDH (1.1.1.27)	Bovine muscle	NAD[+]	Pyruvate	574	2.2	3.6E+11	[10]
L-LactateDH (1.1.1.27)	*Bacillus subtilis*	NAD[+]	Pyruvate	–	–	3.6E+11	[10]
NAD(P)H–FMN Reductase	*Beneckea harveyi* MB-20	NAD[+]	FMN	34.7	–	–	[86]
NAD(P)H–FMN Reductase	*Beneckea harveyi* MB-20	NADP[+]	FMN	81.7	–	–	[86]
Diaphorase	*Clostridium kluyveri*	NAD[+]	2,6-Dichlorophenol-indophenol	1000	–	–	[53]
Diaphorase	*Clostridium kluyveri*	NADP[+]	2,6-Dichlorophenol-indophenol	600	–	–	[53]
Diaphorase	*Clostridium kluyveri*	NAD[+]	2,6-Dichlorophenol-indophenol	3–10	80	–	[71]
Diaphorase	Porcine heart	NAD[+]	K$_3$Fe(CN)$_6$	40	–	–	[50–52]
Diaphorase	Porcine heart	NAD[+]	2,6-Dichlorophenol-indophenol	2–4	1000	–	[71]
LipoamideDH (1.6.4.3)	Porcine heart	NAD[+]	Lipoamide	72.7	13	–	[55]
LipoamideDH (1.6.4.3)	Tolura yeast	NAD[+]	Lipoamide	25–50	7.7	–	[71]

94

NADH Oxidase	*Streptococcus faecalis*	NAD^+	O_2	67.9	–	–	[87]
NADH Oxidase	*Leuconostoc mesenteroides*	NAD^+	O_2	12.9	–	–	[88]
NADH Peroxidase (1.11.1.1)	*Streptococcus faecalis*	NAD^+	H_2O_2	41.5	1520	Irreversible	[10, 300]
Respiratory Chain	Thermophilic bacteria	NAD^+	O_2	~0.1	–	–	[93]
Respiratory Chain	*Escherichia coli*	NAD^+	O_2	~0.1	–	–	[92, 93]

[a] Reagent price in US$.

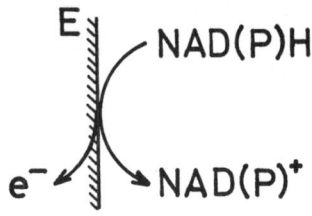

(A) Direct oxidation

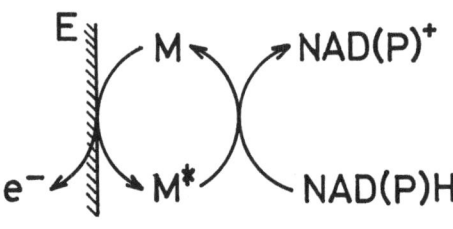

(B) Mediated oxidation

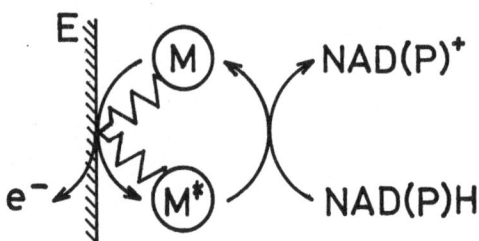

Fig. 2.1. Electrochemical oxidation of NAD(P)H: electrode being represented as E; mediator as M

(C) Oxidation by chemically modified electrode

rapid reoxidation by air oxygen [67]. MANSSON et al. [66] used FMN, FAD, and acriflavin immobilized on Sepharose for the regeneration of NAD(P)$^+$. Acriflavin on epoxy-activated Sepharose 6B were most stably immobilized. Advantages of the chemical method over the enzymatic method are: the wider range of operating conditions, the lower cost [64], and the applicability for both NAD$^+$ and NADP$^+$. A disadvantage is the relatively slow kinetics.

Electrochemical methods for the regeneration of NAD(P)$^+$ are classified into three groups: (1) direct electrode oxidation [72–75], (2) mediated oxidation, and (3) chemically modified electrode approach [76–78]. These are schematically shown in Fig. 2.1. The direct oxidation of NAD(P)H is characterized by a high overpotential: 1.1 V at a carbon and 1.3 V at a platinum electrode for NADH [78]. The mediated oxidation and the chemically modified electrode approach are attempts for the reduction of the overpotential. For more details, see Chap. 1.

Possible enzymatic methods are listed in Table 2.3. Among these, ADH, glutamate dehydrogenase (GlutDH) and LDH are the most important from a practical point of view. The definition of the equilibrium constant in Table 2.3 is the inverse of K_{eq} in Eq. (2.1) in M^{-1}. Equilibria for all the enzyme reactions listed in this table are favorable for the oxidative regeneration of pyridine coenzymes.

ADH again has been most conveniently used because of the low commercial price. The thermodynamic equilibrium of the ADH reaction, in this case, is favorable for the oxidative regeneration of pyridine nucleotide coenzymes. The kinetics is also favorable; the kinetics with yeast enzyme in the oxidative mode is ten times high as in the case of reductive mode.

Various types of glutamate dehydrogenase are available for the regeneration of both NAD^+ and $NADP^+$ with good specific activities at low prices. A problem in this case is a relatively high cost of the substrate, α-ketoglutarate.

LDH has many advantages: favorable equilibrium, commercial availability at a low price and high specific activity. The substrate (pyruvate), however, is moderately expensive.

NAD(P)H-FMN reductase was purified from a marine luminous bacterium [86]. This enzyme may be useful to accelerate the chemical oxidation of NAD(P)H by FMN. DIA and lipoamide dehydrogenase are used together with appropriate electron mediators, which may be regenerated electrochemically.

A NADH oxidase with a substantial specific activity has been discovered [87, 88]. This enzyme works with the least expensive substrate, molecular oxygen from air, but is not very stable. NADH peroxidase [300] was used with hydrogen peroxide as a substrate, which may have a negative effect on the enzyme.

The respiratory chain of *Escherichia coli* was used for the reoxidation of NADH [92]. This system was very instable. To improve the stability, thermophilic bacteria were used. Upon immobilization, the half-life of the oxidizing activity of the respiratory chain was improved [93] but still far from a possible practical use.

2.2.3 Adenosine 5′-Triphosphate (ATP)

No appropriate chemical method is available for the regeneration of ATP from ADP or AMP [94]. The most frequently used and practical method is the enzymatic method listed in Table 2.4. All of the reactions in this table have favorable equilibria for the regeneration of ATP. The equilibrium constant, K_{eq}, in Table 2.4 is defined as:

$$K_{eq} = [ATP][product]/[ADP][substrate] \tag{2.2}$$

for enzymes other than adenylate kinase and

$$K_{eq} = [ADP]^2/[AMP][ATP] \tag{2.3}$$

for adenylate kinase.

Among the enzymes in Table 2.4, acetate kinase (AcK) seems most promising for its ready availability, cost, and the cost of substrate (acetylphosphate) for the phosphorylation. A heat-stable enzyme from a thermophilic bacterium with an excellent specific activity (1719 U/mg) was also purified [97] and is now commer-

Table 2.4. Enzymatic regeneration of ATP

Enzyme (EC)	Origin	Substrate	Specific activity [U/mg]	Price[a] [$/1000U]	Equilibrium constant, K_{eq} [−]	Ref.
Acetate kinase (2.7.2.1)	*Escherichia coli*	Acetyl-phosphate	170	62	125	[10]
Acetate kinase (2.7.2.1)	*Streptococcus faecalis*	Acetyl-phosphate	–	–	125	[10]
Acetate kinase (2.7.2.1)	*Bacillus stearo-thermophilus*	Acetyl-phosphate	1710	260	125	[97]
Adenylate kinase (2.7.4.3)	Rabbit muscle	ATP/AMP	2000	7.6	2.26	[10]
Adenylate kinase (2.7.4.3)	*Bacillus stearo-thermophilus*	ATP/AMP	2000	88.5	2.26	[71]
Carbamate kinase (2.7.2.2)	*Streptococcus faecalis*	Carbamoyl-phosphate	446	84	25	[105]
Carbamate kinase (2.7.2.2)	*Streptococcus* strain D_{10}	Carbamoyl-phosphate	1300	–	25	[10]
Carbamate kinase (2.7.2.2)	*Lactobacillus buchneri* NCDO110	Carbamoyl-phosphate	59.5	–	25	[107]
Pyruvate kinase (2.7.1.40)	Rabbit muscle	Phosphoenol-pyruvate	445	6.3	6450	[108]
Pyruvate kinase (2.7.1.40)	*Bacillus stearo-thermophilus*	Phosphoenol-pyruvate	100–200	50	6450	[71]
Creatine kinase (2.7.3.2)	Rabbit muscle	Phospho-creatine	1880	4.5	1.4E+8	[10]

[a] Reagent price in US$.

cially available. Efficient methods to produce acetylphosphate, the substrate for ATP regeneration, were also investigated [95, 96]. The reaction of AcK may be combined with adenylate kinase (AdK) if the regeneration of ATP from AMP is needed. AdK from a thermophile was also investigated [99] and is commercially available. Attempts to improve the stability of these enzymes by immobilization were also made [98–102]. NAKAJIMA et al. immobilized AcK from *Bacillus stearothermophilus* on Sepharose resin [98]. The immobilized enzyme retained more than 80% of the initial activity after one month continuous operation. Coimmobilization of AcK and AdK both from *Bacillus stearothermophilus* on Sepharose 4B was attempted [99]. The immobilized enzyme system converted more than 99% of AMP to ATP over a period of 6 days.

Carbamate kinase (CK) also has been well studied [103–107]. This enzyme has a good specific activity. The instability of the substrate, carbamoylphosphate, in an aqueous solution is a major problem. MARSHALL [106] immobilized CK from *Streptococcus faecalis* on alkylamine glass. The immobilized enzyme lost only 16% of the initial activity after continuous operation for 14 days. Instead of carbamoylphosphate as a substrate, he used KH_2PO_4 and KOCN, which slowly

Table 2.5. ATP regeneration through energy-transfer pathways

Energy transfer system	Origin	State of system	Substrate	Specific activity [U/mg]	Ref.
Glycolytic pathway	*Saccharomyces cerevisiae*	Crude enzyme	Glucose, adenosine	1.1E−4/dry cell mass	[113–116]
Glycolytic pathway	*Brevibacterium ammoniagenes*	Permealized cell	Glucose, adenosine	1.5E−4/dry cell mass	[117]
Glycolytic pathway	*Saccharomyces cerevisiae*	Gene recombinant, dried cell	Glucose, adeninosine	2.7E−3/dry cell mass	[118]
Oxidative phosphorylation	Bovine heart	Mitochondrial electron transfer particles	(Succinate)	2.7E−3/protein	[119]
Oxidative phosphorylation	*Candida boidinii*	Protoplast	Methanol or formate	0.024/protein	[120]
Photophosphorylation	Thermophilic blue-green algae	Whole cell	Light	0.63/chlorophyl	[121]
Photophosphorylation	*Rhodospirillum rubrum*	Chromatophore	Light	0.62/chlorophyl	[122]
Photophosphorylation	Lettuce	Thylakoid	Light	3.9/chlorophyl	[124]
Photophosphorylation	Lettuce	Immobilized thylakoid	Light	0.37/chlorophyl	[124]
Photophosphorylation	*Rhodopseudomonas capsulata*	Chromatophore	Light	9.3/chlorophyl	[125–127]
Photophosphorylation	*Rhodopseudomonas capsulata*	Immobilized chromatophore	Light	6.2/chlorophyl	[125–127]

reacted to form carbamoylphosphate. The carbamoylphosphate formed was quickly consumed in the CK reaction. Thus the instability problem of the substrate could be circumvented.

HIRSCHBEIN et al. [109] studied a convenient method for synthesizing phosphoenolpyruvate (PEP). They pointed out possible advantages of the pyruvate kinase (PK)/PEP system over the AcK/acetylphosphate system: PEP is more easily synthesized on the laboratory scale, more stable in aqueous solution, and stronger as a phosphorylating agent. A disadvantage is the higher cost of the starting material (pyruvic acid) for the synthesis of PEP.

Creatine kinase [111] has an advantage in the stability of the substrate (phosphocreatine), the cost of which, however, is high. Creatine kinase immobilized in a polyamphoteric gel continued operating for 10 days with a conversion of 85% [112].

Besides the single-enzyme method for ATP regeneration, methods with energy-transfer pathways inside of cells were also investigated as listed in Table 2.5. ASADA et al. [113–116] used the alcohol fermentation pathway and kinases of yeast for the continuous regeneration of ATP from adenosine. A crude enzyme extract from *Saccharomyces cerevisiae* was immobilized in a collodion membrane reactor system. A steady-state activity with a yield over 70% for ATP was maintained for 13 days [116]. Recombinant DNA techniques were applied to intensify the phosphorylating activity in the glycolytic pathway [118]. Yeast cells harboring the hybrid plasmid carrying the genes of hexokinase and glucokinase showed an increased ATP-producing activity when used after being dried. Resting cells of *Brevibacterium ammoniagenes* were used to produce ATP from adenine [117]. Usage of the cationic detergent, polyoxyethylenestearylamine, allowed ATP to permeate through the cell membrane.

The oxidative phosphorylation system as an ATP regenerator was investigated by several groups. MATSUOKA et al. [119] tried to immobilize beef heart mitochondrial electron transfer particles in agar gel. The immobilized system kept 80% of the initial activity or more after being used five times. TANI et al. [120] used a protoplast of *Candida boidinii,* which could produce ATP from AMP upon addition of methanol or formate. Participation of the substrate level phosphorylation in the ATP formation could be ignored.

ATP could also be regenerated by photophosphorylation [121–127]. COCQUEMPOT et al. [124] used lettuce thylakoids and chromatophores from *Rhodopseudomonas capsulata.* The half-lives of the activity were 30 min and 1 day without immobilization, respectively, and were improved to 60 min and 8 days with immobilization, respectively. SMEDS et al. [122] used chromatophores of *Rhodospirillum rubrum* in an aqueous two-phase system of polyethyleneglycol and dextran. SAWA et al. [121] used thermophilic blue-green algae, the half-life of which was 3 days. All of these challenging approaches with energy-transfer pathways in entire cells seem to require further im-provement in their activity and stability before being practically applied.

2.3 Enzyme Processes with Coenzyme Cycling

2.3.1 Process with Regeneration of NAD(P)H

An enzymatic process involving NAD(P)H regeneration is stereospecific reduction, which is synthetically difficult. Investigations on this process are classified into three categories depending on the method of coenzyme regeneration: chemical, substrate conjugation, and enzymatic. These are listed in Tables 2.6–2.8. In these tables, the NAD cycling number and cycling rate are indices for the effective usage of coenzyme. For example, when the NAD cycling number is R_N, R_N molecules of product are produced from one molecule of NAD. This means that the cost of NAD is reduced to $1/R_N$. Productivity is an index for the performance of a reactor expressed in mol/l/h. The higher the productivity, the smaller the reactor volume will be. Hydroxy acids, alcohols, amino acids, isotope-labelled compounds, and pharmaceuticals are produced with NAD(P)H recycling.

2.3.1.1 Hydroxy Acids

Probably because of the easy commercial availability of LDH, the production of lactate from pyruvate with NADH cycling has been most extensively studied

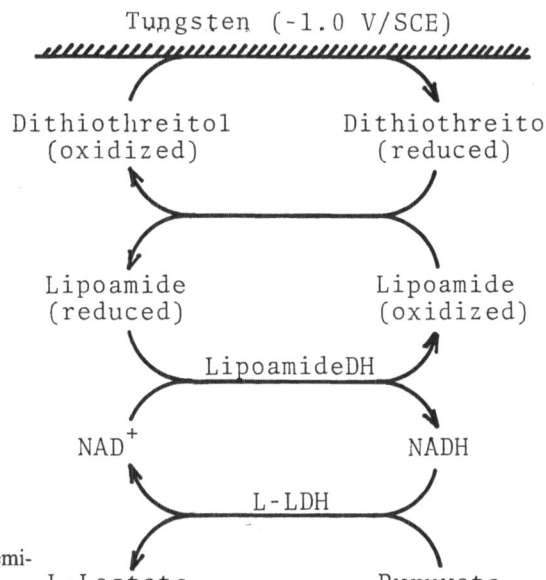

Tungsten (-1.0 V/SCE)

Dithiothreitol (oxidized)   Dithiothreitol (reduced)

Lipoamide (reduced)   Lipoamide (oxidized)

LipoamideDH

NAD$^+$ NADH

L-LDH

L-Lactate   Pyruvate

Fig. 2.2. LDH reaction with electrochemical regeneration of NADH

rather from an academic interest as a convenient model reaction with coenzyme cycling. However, optically pure L-lactate and especially D-lactate may be of some practical interest. SHAKED et al. [128] combined reactions of LDH and lipoamide dehydrogenase with the electrode reaction. NADH consumed in the LDH reaction to produce lactate was regenerated back from NAD$^+$ by reduced lipoamide, which was regenerated from oxidized lipoamide by reduced dithiothreitol. The reduced dithiothreitol finally is regenerated electrochemically at a tungsten electrode as shown in Fig. 2.2. The transhydrogenation between NAD$^+$ and the reduced lipoamide was catalyzed by lipoamide dehydrogenase. DiCOSMO et al. [129] studied the same type system with methyl viologen as the only mediator between the enzymatic reaction for the NADH regeneration and the electrode reaction. For the transhydrogenation between the coenzyme and the mediator, either lipoamide dehydrogenase or ferredoxin reductase was used. The results of all these systems were almost the same for coenzyme cycling number and productivity. The chemical step and the electrochemical step seem rate limiting. The overall cycling rate of NADH was lower than in the case of the enzymatic method for coenzyme regeneration.

A conjugated system of LDH with yeast alcohol dehydrogenase (YADH) has been well studied by many investigators. WYKES et al. [144] immobilized LDH and YADH on the same support of DEAE-cellulose. In this system, formylpolyethyleneimine-bound NAD was an effective coenzyme. MORIKAWA et al. [145] entrapped dextran-bound NAD together with LDH and YADH in a collagen matrix, which worked as a self-contained system with no need for external addition of NAD. MANSSON et al. [234] prepared a conjugated compound of NAD with horse liver alcohol dehydrogenase (HLADH) in the attempt to make NAD an artificial prosthetic group of HLADH. This compound was active with free LDH

Table 2.6. Enzyme processes with chemical and electrochemical regeneration of NAD(P)H

Main product	Enzyme for main reaction	Method for coenzyme regeneration	Main substrate	Cosubstrate	NAD cycling		Productivity [mol/l/h]	Remarks[a]	Ref.
					Number [-]	Rate [h⁻¹]			
L-Lactate	L-LDH	LipoamideDH/Lipoamide/DTT/Tungsten (-1.0V SCE)	Pyruvate	H^+, e^-	920	~ 11	3.1E-3	B, IME	[128]
L-Lactate	L-LDH	LipoamideDH/MV^{2+}/Tungsten (-0.7V SCE)	Pyruvate	e^-	940	~ 4.4	6.6E-4	B, IME	[129]
L-Lactate	L-LDH	Ferredoxin Rase/MV^{2+}/Tungsten (-0.7V SCE)	Pyruvate	e^-	900	~ 2.7	5.4E-4	B, IME, NADPH	[129]
Cyclohexanol	HLADH	[Rh(bpy)$_2$]⁺/Graphite (-0.9V Ag/AgCl)	Cyclohexanone	e^-	2.9	–	–	B	[131]
Cyclohexanol	HLADH	Chemical	Cyclohexanone	$Na_2S_2O_4$	105	–	–	B	[62]
L-Carnitine	CarnitineDH	Chemical	3-Dehydrocarnitine	$Na_2S_2O_4$	~ 22.3	~ 1.3	1.3E-3	B	[23]
2-Butanol	ADH	FDR/MV^{2+}/Ru(bby)$_3^{2+}$	2-Butanone	$h\nu$, $(NH_4)_3$ EDTA	40	1	1E-3	B, NADPH	[132]
2-Norbornanol	HLADH	LDH/Pyruvate/bis-(phosphine)rhodium/H_2	2-Norbornanone	H_2	~ 49	~ 0.26	7.8E-5	B	[130]
Cyclohexanol	HLADH	LDH/Pyruvate/bis-(phosphine)rhodium/H_2	Cyclohexanone	H_2	~ 50	~ 0.26	7.8E-5	B	[130]
L-Glutamate	GlutamateDH	FDR/MV^{2+}/Tungsten (-0.7V SCE)	NH_4^+, α-Ketoglutarate	e^-	1000	~ 6.0	1.2E-3	B, IME	[129]
Malate	MDH	DIA/MV^{2+}/Glassy carbon (-0.9V Ag/AgCl)	Oxaloacetate	e^-	~970	~190	9.5E-3	B	[133]

[a] Abbreviations: B, batchwise operation; IME, immobilized enzyme.

Table 2.7. Enzyme processes with substrate conjugation method for NAD(P)H regeneration

Main product	Enzyme for main reaction	Main substrate	Cosubstrate	NAD cycling		Productivity [mol/l/h]	Remarks[a]	Ref.
				Number [−]	Rate [h^{-1}]			
Propanediol	HLADH	Lactoaldehyde	Ethanol	~53	~212	0.148	B	[134]
Benzylalcohol	HLADH	Benzaldehyde	Ethanol	~84	~2400	0.024	C, IME	[135]
Propylalcohol	YADH	Propionaldehyde	Ethanol	~140000	~240	0.024	C, IME	[141]
Propanediol	HLADH	Lactoaldehyde	Ethanol	–	40000	1.58E−3	B, ADH–NAD	[233]
Thiopyranols	HLADH	Thiopyranones	Ethanol	~3.4	~0.06	1.38E−4	B	[136]
Hydroxyketones	HLADH	Decalindiones	Ethanol	–	–	–	B	[138]
3-Hydroxybutyrate	HLADH	Ethylacetoacetate	Ethanol	~8.2	~0.05	6.9E−5	B	[140]
3-Hydroxybutyrate	YADH	Ethylacetoacetate	Ethanol	~6.1	~0.04	5.5E−5	B	[140]
(−)-7-Hydroxy-cis-decalin-2-one	HLADH	cis-Decalin-2.7-dione	Ethanol	~4.0	~0.003	7.5E−6	B	[139]
R(+)-2-pentanol	ADH (Thermo-anaerobium brochii)	2-Pentanone	2-Propanol	20000	500	2.5E−2	B, NADH or NADPH	[137]

[a] Abbreviations: B, batchwise operation; C, continuous operation; IME, immobilized enzyme; ADH–NAD, ADH–NAD conjugate was used.

Table 2.8. Enzyme processes with enzymatic regeneration of NAD(P)H

Main product	Enzyme for main reaction	Method for coenzyme regeneration	Main substrate	Cosubstrate	NAD cycling Number [−]	NAD cycling Rate [h^{-1}]	Productivity^a [mol/l/h]	Remarks	Ref.
Ethanol-1-*d*	YADH	GlucoseDH	CH$_3$CDO	Glucose	~5307	~786	0.149	B	[142]
Alanine	AlaDH	GalactoseDH	Pyruvate, NH$_4^+$	Galactose	>90	14	2.1E−3	C, IMC, IME	[143]
Alanine	AlaDH	LDH	Pyruvate, NH$_4^+$	Lactate	>180	33	5.94E−3	C, IMC, IME	[143]
Lactate	LDH	YADH	Pyruvate	Ethanol	>1300	< 9.8	2.74E−3	C, IMC, IME	[29]
Lactate	LDH	YADH	Pyruvate	Ethanol	>120	~ 40	4E−4	B	[144]
Lactate	LDH	YADH	Pyruvate	Ethanol	>15	~ 25	2.5E−4	C, IMC, IME	[144]
Lactate	LDH	YADH	Pyruvate	Ethanol	–	29	9.3E−5	C, IMC, IME	[145]
Lactate	LDH	YADH	Pyruvate	Ethanol	7500	7550	2.11E−3	C, IME	[146]
Lactate	LDH	YADH	Pyruvate	Ethanol	412000	820	4.1E−2	C, IME	[147]
Lactate	T-LDH	T-MDH	Pyruvate	Malate	>120	0.6	6E−4	C, IMC, IME	[177]
L-Carnitine	CarnitineDH	YADH	3-Dehydrocarnitine	Ethanol	89.3	1.9	1.9E−3	B	[23]
L-Carnitine	CarnitineDH	YADH + AldDH	3-Dehydrocarnitine	Ethanol	120	2.6	2.6E−3	B	[23]
L-Carnitine	CarnitineDH	GlucoseDH	3-Dehydrocarnitine	Glucose	530	7.4	7.4E−3	B	[23]
Lactate	LDH	HLADH–NAD	Pyruvate	Ethanol	>40	300	3.3E−4	B, ADH–NAD	[234]
L-Malate	MDH	FormateDH	Oxaloacetate	Formate	~ 46	0.69	2.1E−3	C, IMC, IME	[235]
L-Malate	T-MDH	FormateDH	Oxaloacetate	Formate	260	0.52	1.3E−3	C, IMC, IME	[236]
Epoxide	Methane mono-oxygenase	FormateDH etc.	Alkene	Formate etc.	–	–	–	B	[150]
Alcohol, Epoxide	Methane mono-Oxygenase	FormateDH or sec ADH	Alkane etc.	Formate	–	–	–	B	[151]
Diastereomeric 4-methyl-L-glutamate	GlutamateDH	YADH	2-Keto-4-methyl-glutamate	Ethanol	20	~ 0.10	7.5E−4	B	[152]
Xylitol	Xylose reductase	F$_{420}$–NADP reductase Hydrogenase	Xylose	H$_2$	~ 16.5	~ 4.1	1.1E−2	C, IMC, IME, NADPH	[153]
12-Ketocheno-deoxy-cholic	3-α-Hydroxy-steroidDH	FormateDH	Dehydrocholic acid	Formate	1200	14.3	5,4E−4	B	[154]

acid	7-α-Hydroxy-steroidDH	NAD-reductase									
Alanine	AlanineDH		Pyruvate	H₂	~	6.3	~	1.25	1E−3	B, IME, at 100 atm	[155]
Malate	MalateDH	YADH	Oxaloacetate	Ethanol	>	20	~	20	3.9E−4	B, IME	[156]
Malate	MalateDH	YADH	Oxaloacetate	Ethanol	>	72	~	72	1.3E−3	B, IMC, IME	[157]
Malate	MalateDH	YADH	Oxaloacetate	Ethanol	>	100	~	15	8.1E−3	C, IMC, IME	[157]
Malate	MalateDH	YADH	Oxaloacetate	Ethanol	>	4.9	~	4.9	1.3E−3	B, IMC, IME	[158]
Malate	MalateDH	YADH	Oxaloacetate	Ethanol	>	19.0	~	6.4	4.8E−3	C, IMC, IME	[158]
Glutamate	GlutamateDH	GlucoseDH	α-Ketoglutarate, NH₃	Glucose	−	−		−	−	B, IME, NADH or NADPH	[159]
Glutamate	GlutamateDH	YADH	α-Ketoglutarate, NH₃	Ethanol		40		13.3	8.3E−4	B, IMC, IME, NADPH	[160]
Glutamate	GlutamateDH	YADH	α-Ketoglutarate, NH₃	Ethanol		40		13.3	8.3E−4	B, IMC, IME	[161]
D(−)-β-Hydroxybutyrate	D(−)-β-HydroxybutylateDH	D-GalactoseDH	Acetoacetate	D(+)-Galactose		12.8		0.27	6.8E−4	B	[163]
D(−)-β-Hydroxybutyrate	D(−)-β-HydroxybutylateDH	D-GalactoseDH	Acetoacetate	D(+)-Galactose		−		−	−	B	[164]
D-Lactic acid	D-LacticDH	GlucoseDH (Bacillus cereus)	Pyruvate	Glucose		40000		280	2.8E−3	B, IME	[170]
Ethyl-(R)-4-chloro-3-hydroxybutanoate	HLADH or TADH	GlucoseDH (Bacillus cereus)	Pyruvate	Glucose		72		7.2	7.2E−4	B, IMC, IME, NADH or NADPH	[170]
(R)-2,2,2-Trifluoro-1-phenylethanol	TADH	GlucoseDH (Bacillus cereus)	Pyruvate	Glucose		90		9.0	9.0E−4	B, IME, NADH or NADPH	[170]
Ethyl-(S)-3-hydroxyvalerate	HLADH or TADH	GlucoseDH (Bacillus cereus)	Pyruvate	Glucose		90		9.0	9.0E−4	B, IME, NADH or NADPH	[170]
(S)-Lactaldehyde dimethyl acetal	YADH or HLADH or TADH	Glucose DH (Bacillus cereus)	Pyruvate	Glucose		90		9.0	9.0E−4	B, IME, NADH or NADPH	[170]
(S)-3-Hydroxybutanol dimethyl acetal	YADH or HLADH or TADH	GlucoseDH (Bacillus cereus)	Pyruvate	Glucose		90		9.0	9.0E−4	B, IME, NADH or NADPH	[170]

Table 2.8 (continued)

Main product	Enzyme for main reaction	Method for coenzyme regeneration	Main substrate	Cosubstrate	NAD cycling Number [−]	NAD cycling Rate [h⁻¹]	Productivity^a [mol/l/h]	Remarks	Ref.
(S)-Benzyl-α-d_1-alcohol	HLADH	G6PDH	Benzaldehyde-α-d_1	G6S	1000	5.2	8.3E−4	B, IME, Free CoE	[165]
threo-D_s(+)-Isocitrate	IsocitrateDH	G6PDH	α-Ketoglutarate	G6S	1500	12.5	1.3E−3	B, IME, NADPH	[165]
D-Lactic acid	D-LDH	G6PDH	Pyruvate	G6P	1200	12.5	6.3E−3	B, IME	[166]
threo-D_s(+)-Isocitrate	IsocitrateDH	G6PDH	CO_2, α-Keto-glutarate	G6P	1020	10.6	2.1E−3	B, IME	[166]
(S)-Benzyl-α-d_1-alcohol	HLADH	G6PDH	Benzaldehyde-α-d	G6P	1000	20.8	1.0E−2	B, IME	[166]
D-Lactic acid	D-LDH	G6PDH, Phospho-fructomutase	Pyruvate	Fructose-6-phosphate	1650	34.3	4.3E−3	B, IME	[166]
L-Lactate	L-LDH	DIA or ADH or GlutamateDH/6-Phospho-gluconateDH and Phosphoribo-isomerase	Pyruvate	6-Phospho-gluconate	1000	5.2	1.0E−3	B, IME, NADH and NADPH	[167]
threo-D_s(+)-Isocitrate	IsocitrateDH	DIA/Glutamate-DH/FDH	α-Ketoglutarate, CO_2	Formate	1000	3.8	7.6E−4	B, IME, NADH and NADPH	[167]
threo-D_s(+)-Isocitrate	IsocitrateDH	ADH/FDH	α-Ketoglutarate, CO_2	Formate	1000	~ 3.8	7.6E−4	B, IME, NADH and NADPH	[167]
L-Chlorlactate	L-LDH	G6PDH	Chloropyruvate	G6P	394	2.35	5.0E−4	B, IME	[168]
D-Chlorlactate	D-LDH	G6PDH	Chloropyruvate	G6P	400	2.38	6.0E−4	B, IME	[168]
(R)-Trifluoro-ethanol-1-d_1	HLADH	FDH	Trifluoroaldehyde dihydrate	DCO_2Na	600	2.5	2.5E−4	B, IME	[169]
L-Glutamic-α-d_1 acid	GlutamateDH	AldehydeDH	α-Ketoglutarate, NH_3	Glyco-aldehyde-1,2,2-d_3	960	20.9	2.1E−3	B, IME	[169]

L-Glutamic-α-d_1 acid	GlutamateDH	YADH	α-Ketoglutarate, NH$_3$	Ethanol-1,1-d_2	501	10.4	1.0E−3	B, IME	[169]
Alanine	AlanineDH	FormateDH	Pyruvate, NH$_3$	Formate	>4780	~ 96	0.35	C, IMC, IME	[172]
D-Phenyllactate	D-LDH	FormateDH	Phenylpyruvate	Formate	~ 830	~ 5.2	5.2E−3	C, IMC, IME	[172]
L-Alanine	AlanineDH	D- and L-LDH	DL-Lactate	NH$_3$	19 200	–	–	C, IMC, IME	[175]
L-2-Hydroxy-isocaproate	L-2-Hydroxy-isocaproateDH	FormateDH	2-Ketoisocaproate	Formate	69 400	~320	0.16	C, IMC, IME	[174]
D-2-Hydroxy-isocaproate	D-2-Hydroxy-isocaproateDH	FormateDH	2-Ketoisocaproate	Formate	22 800	~190	0.19	C, IMC, IME	[174]
L-Leucine	L-LeucineDH	FormateDH	α-Ketoisocaproate, NH$_3$	Formate	~5110	~ 4.4	1.3E−2	C, IMC, IME	[171]
L-Leucine	L-LeucineDH	FormateDH	α-Ketoisocaproate, NH$_3$	Formate	18 200	~ 32	1.8E−2	C, IMC, IME	[173]
L-Leucine	L-LeucineDH (*Bacillus stearothermophilus*)	FormateDH	α-Ketoisocaproate, NH$_3$	Formate	50 000	~ 72	7.2E−2	C, IMC, IME	[176]

[a] Abbreviations: B, batchwise operation; C, continuous operation; IMC, immobilized coenzyme; IME immobilized enzyme.

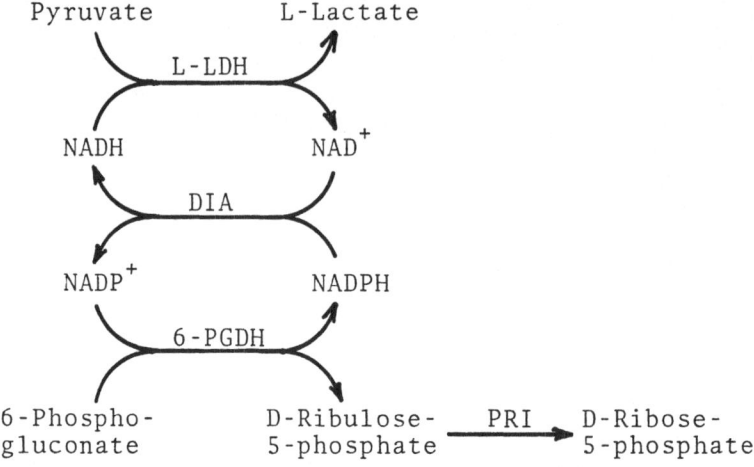

Fig. 2.3. LDH reaction with NADH regeneration involving transhydrogenation between NADPH and NAD$^+$ (Abbreviations: LDH, lactate dehydrogenase; DIA, diaphorase; 6-PGDH, 6-phosphogluconate dehydrogenase; PRI, phosphoriboisomerase)

in the presence of pyruvate and ethanol. The NAD bound on the HLADH protein was cycled effectively, giving a cycling rate of 300 h^{-1}. WONG et al. [167] used LDH together with a combined system of DIA/6-phosphogluconic dehydrogenase (6-PGDH)/phosphoriboisomerase (PRI) for NADH regeneration (Fig. 2.3). As for coenzyme regeneration, systems with ADH/6-PGDH/PRI and glutamate dehydrogenase/6-PGDH/PRI were also used. These systems involve a transhydrogenation reaction between NAD and NADP, that makes selection of the regeneration system of pyridine nucleotides more flexible.

In attempts to construct a continuously working system, YAMAZAKI et al. [29] prepared polyethyleneimine-bound NAD, which was physically immobilized in a membrane reactor system along with LDH and YADH. This constituted a self-contained system, which functioned for 122 h with a 31% loss of overall activity under the continuous feed of the substrates (pyruvate and ethanol). The cycling number of NADH was at least 1300, i.e., one molecule of NAD produced 1300 molecules of lactate. HAYAKAWA et al. [177] also used a membrane reactor containing polyethylene glycol-bound NAD, thermostable LDH, and malate dehydrogenase (MDH). In a continuous operation, the half-life of the polyethylene glycol-bound NAD, however, was the shortest (5.7 days at 30 °C) among the immobilized catalysts including enzymes and coenzyme.

MIYAWAKI et al. [146] employed a system with LDH and YADH coimmobilized at a high concentration. To this system, substrates and free NAD at a low concentration level were fed continuously. When the NAD concentration in the feed solution was lower than that of the immobilized enzymes on the molecular basis, NAD fed was locally concentrated around the enzymes through the affinity effect. This may be considered as a "dynamic immobilization" of coenzyme. As a result, the NAD cycling number was as high as 7500. This idea was extended to propose an affinity chromatographic reactor [147]. In this case, the conjugated enzymes, LDH and YADH, were coimmobilized at a high concentration level in a column-type reactor. The coenzyme needed, NAD, was fed as a pulse for a short time, while other substrates were fed continuously. Because of the strong affinity interaction with the immobilized enzymes, the retention time of the NAD pulse was very long and the cycling number of NAD was as high as 412 000. More details on this will be given in Sect. 2.5.3.

D-Lactate is a useful synthon for the preparation of chiral substances. To prepare this, D-LDH was used together with glucose-6-phosphate dehydrogenase (G6PDH) or a combined system of G6PDH/phosphofructomutase. The cycling number of NAD was between 1200 and 1700 [166]. Later, the regenerating system of coenzyme was replaced with glucose dehydrogenase [170]. A high concentration of enzymes and substrate (pyruvate) and a low concentration of NAD gave a cycling number of coenzyme as high as 40 000 with a conversion of 90%.

D-Chlorolactate and its L-isomer, useful starting materials for more complex enantioselective syntheses, were produced from chloropyruvate with a relatively low cycling number of coenzyme around 400 [168]. This is because the substrate (chloropyruvate) was reactive toward the enzymes and the lifetime of NAD was relatively short in this system.

D-Phenyllactate was continuously formed from phenylpyruvate by use of a combined system of D-LDH, FDH, and polymer-bound NAD, all of which were immobilized in a membrane reactor system [172].

CAMPBELL et al. [156] immobilized MDH and YADH in microcapsules to produce malate from oxaloacetate with NAD cycling. NAD was externally added and ethanol was used as a cosubstrate for regeneration of NADH. Later, they prepared dextran-bound NAD, which was coimmobilized with MDH and YADH in microcapsules [157]. This self-contained system produced malate continuously when oxaloacetate and ethanol was fed. The cycling number of NAD was more than 100. To improve the mechanical strength of the microcapsule, crosslinking with glutaraldehyde was attempted [158]. This, however, caused a loss in enzyme activity. The cycling number of NAD in the continuous operation for 3 h was only about 19.

YAMAZAKI et al. [235] prepared a polymerizable NAD derivative, N^6-[N-[N-2(hydroxy-3-methacrylamidopropyl) carbamoyl-methyl] -carbamoylmethyl]-NAD, which could form a gel entrapping pig-heart MDH and FDH. A column containing this gel continuously produced malate from oxaloacetate with no external feed of NAD. However, 90% of the activity was lost in 3 days because of the insufficient stability of the pig-heart MDH. The NAD cycling number was 46 in this case. Later, pig-heart MDH was replaced with MDH from *Thermus thermophilus* [236]. The overall activity continued for more than 3 weeks and the cycling number of NAD was improved to be 260. These attempts for the construction of a self-contained gel immobilizing coenzyme together with enzymes seem very attractive if the activity and the stability of the immobilized biocatalysts are improved further. Compared with the self-contained membrane reactor system, a drawback of the self-contained gel system exists in the impossibility of external resupply of immobilized enzymes and coenzyme when some of them are inactivated.

MAEDA et al. [133] attempted to combine the MDH reaction with an electro-chemical regeneration technique of NADH. In an electrochemical reaction cell containing MDH, DIA, NAD, oxaloacetate, and a mediator (methyl viologen), a potential of -0.9 V (vs. Ag/AgCl) was impressed on the working electrode, glassy carbon beads. The NADH consumed in the MDH reaction was regener-ated electrochemically. The cycling number of NAD was 970 and the cycling rate of NAD was 190 h^{-1}. This value seems quite high as a process involving an elec-trode reaction and is probably due to the large surface area of the beads elec-trode.

Enzymatic production of optically active 2-hydroxyisocaproate was carried out in a membrane reactor system containing L- or D-hydroxyisocaproate dehy-drogenase, FDH, and polyethylene glycol-bound NAD [174]. In a continuous op-eration, the deactivation rate of polyethylene glycol-bound NAD was about 9%/day. By choosing appropriate operating conditions, the NAD cycling numbers were quite satisfactory; 69 400 and 22 800 for L- and D-hydroxyisoca-proate dehydrogenase, respectively. The productivity was also good, indicating the possibility for practical application of this method.

2.3.1.2 Alcohols

For the production of alcohols, the substrate conjugation method is available by the effective use of the loose substrate specificity of alcohol dehydro-genases. Under appropriate operating conditions, this method, in principle, should give the highest cycling number and cycling rate of NAD because NAD does not need to dissociate from the apoenzyme in the regeneration cycle.

Lactoaldehyde was converted to propanediol catalyzed by HLADH with etha-nol as a cosubstrate for the regeneration of NADH [27]. The cycling number and the cycling rate of NAD were 53 and 212 h^{-1}, respectively. MANSSON et al. [233] prepared a NAD-HLADH conjugate, which was tested in a substrate conjuga-tion system of lactoaldehyde/ethanol. NAD bound on HLADH effectively cycled to produce propanediol with the extremely high NAD cycling rate of 40 000 h^{-1}. The NAD concentration was very low (4×10^{-8} M) in this case. This made NAD the limiting component in the overall process. Benzaldehyde was converted to benzylalcohol continuously with immobilized HLADH in an ultrafiltration hol-low fiber [135]. Propionaldehyde was continuously converted to propylalcohol in an affinity chromatographic reactor with immobilized YADH [141]. NAD cycling number was as high as 140 000.

2-Substituted thiopyran-4-ones were converted to thiopyranols in the HLADH-catalyzed reaction with NAD recycling. The product, thiopyranols, could be further transformed to enantiometrically pure acyclic secondary alco-hols [136]. *Cis*-Decalindione and its *trans*-isomer were converted by HLADH to enantiometrically pure hydroxyketones, which was chemically converted to (+)-4-twistanone [138, 139].

Transformation of ethylacetoacetate to 3-hydroxybutyrate was catalyzed by HLADH or YADH. 3-Hydroxybutyrate is a useful chiral synthon for further de-rivation to pharmaceuticals and agricultural chemicals. As a regeneration system for NADH, ADH-catalyzed substrate coupling with ethanol as a cosubstrate

110

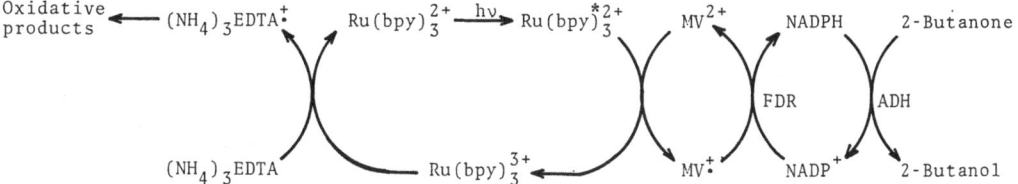

Fig. 2.4. Transformation of 2-butanone to 2-butanol catalyzed by ADH combined with a photosensitized method for NADPH regeneration. (Abbreviation: ADH, alcohol dehydrogenase; FDR, ferredoxin-NADP$^+$ reductase; MV^{2+}, methylviologen)

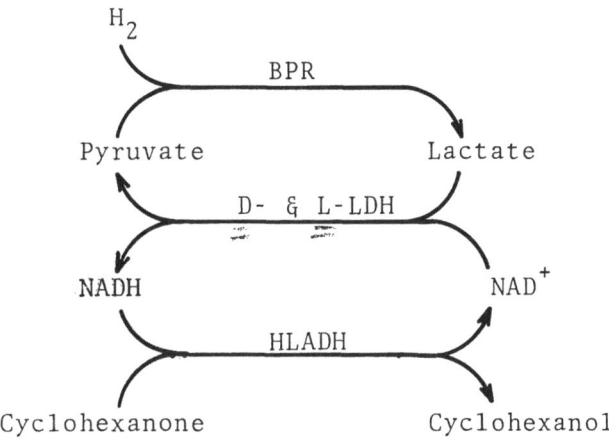

Fig. 2.5. Reduction of cyclo-hexanone by using a hybrid organometallic-enzymatic catalyst. (Abbreviations: LDH, lactate dehydrogenase; HLADH, horse liver alcohol dehydrogenase)

[140] or a D-galactose dehydrogenase/D-galactose system [134] were employed. 2-Pentanol was produced in a substrate conjugation system of ADH/2-pentanone/2-propanol [137].

MANDLER et al. [132] transformed 2-butanone to (−)-2-butanol with an optical purity of 100% by using an ADH-catalyzed reaction coupled with a photosensitized chemical method for NADH regeneration as shown in Fig. 2.4. To convert cyclohexanone to cyclohexanol, WIENCAMP et al. [131] employed a combined system of HLADH/Rh(bpy)$_2^+$/graphite electrode, which was driven electrochemically. Passivation of the graphite cathode caused a small cycling number of NAD (2.9). JONES et al. [62] converted cyclohexanone to cyclohexanol in an HLADH-catalyzed reaction conjugated with direct chemical regeneration of NADH by sodium dithionite, a strong reducing agent. ABRIL et al. [130] obtained cyclohexanol and 2-norbornanol in the reaction catalyzed by HLADH combined with a LDH/lactate/bis(phosphine)rhodium system for NADH regeneration (Fig. 2.5). Pyruvate was reduced back to lactate by bis(phosphine)rhodium and dihydrogen as an ultimate electron donor. Problems in this unique method were slow kinetics and short life-time of the rhodium catalyst.

PATEL et al. [151] used methane monooxygenase conjugated with FDH or secondary ADH to produce alcohols from alkanes. KITPREECHAVANICHI et al. [153] used a xylose reductase/F_{420}-NADP oxidoreductase/hydrogenase system to produce xylitol from xylose and hydrogen.

2.3.1.3 Amino Acids

Amino acids can be produced from corresponding ketoacids through the reductive amination by dehydrogenases. The continuous production of alanine was attempted in conjugated systems of alanine dehydrogenase (AlaDH)/galactose dehydrogenase or AlaDH/LDH with dextran-bound NAD in a membrane reactor as shown in Fig. 2.6. The cosubstrate for NADH regeneration was galactose in the former case and lactate in the latter case [143].

In the latter case, not only NAD but also pyruvate, the intermediate product, was recycled in the reaction. WANDREY et al. [172] employed a conjugated system of AlaDH/FDH with polyethylene glycol-bound NAD (PEG-NAD) to produce alanine. The advantage of FDH as a regenerator of NADH is the generation of an easily separable inert product, carbon dioxide. A disadvantage is its relatively slow specific activity. For the production of alanine, a conjugated system of AlaDH/D,L-LDH with PEG-NAD was also employed [175]. In this case, the only substrate other than ammonia, DL-lactate, is inexpensive and the NAD cycling number was as high as 19 200. These facts show the high potential for the practicability of this process. MATSUNAGA et al. [155] used AlaDH together with hy-

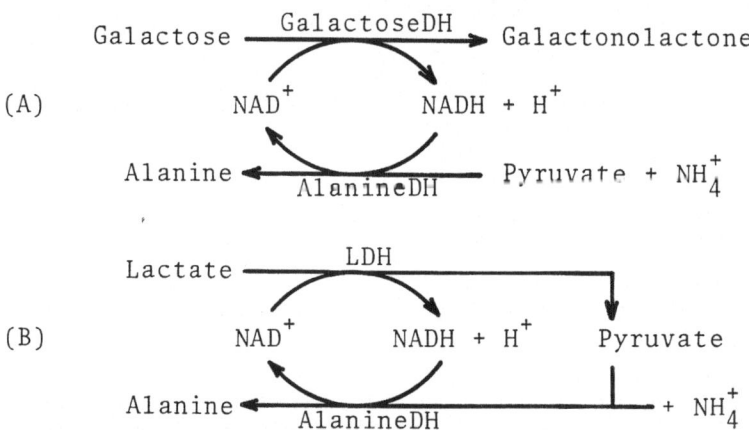

Fig. 2.6. Production of alanine with NADH cycling

112

drogenase in whole cells of *Clostridium butyricum* to produce alanine in the presence of hydrogen at a pressure of 100 atm. The NADH regeneration rate increased with an increase in the pressure of hydrogen.

Ketoglutarate with ammonia was converted to glutamate in conjugated systems of glutamate dehydrogenase/GDH [159] or glutamate dehydrogenase/YADH [160] with NADP cycling. The purpose of these processes is to remove urea or ammonia from body fluids or dialysate from artificial kidney. Glutamate production with cycling of NADP in a conjugated system of glutamate dehydrogenase/ferredoxin reductase/methyl viologen/tungsten electrode (-0.7 V vs. SCE) was also attempted [129].

WANDREY et al. [171, 173] extended their approach for L-alanine production to produce L-leucine continuously from α-ketoisocaproate and NH_4^+ in a conjugated system of leucine dehydrogenase/FDH together with PEG-NAD immobilized in a membrane reactor system. The NAD cycling number was 18 200. Later, they used L-leucine dehydrogenase from *Bacillus stearothermophilus* to improve the operational stability [176]. Here the NAD cycling number was improved to 50 000, which again seems to show the possible practicability of this approach.

TAPIE et al. [152] used a glutamate dehydrogenase/YADH system to produce diastereomeric 4-methyl-L-glutamic acid, which is used in the study of the stereochemistry and reaction mechanism of some enzymes.

2.3.1.4 Isotope-Labeled Compounds

A pure enantiomorph of ethanol-1-d was synthesized from CH_3CDO in a YADH/GDH system with a NAD cycling number of 5300 [142]. (S)-Benzyl-α-d_1 alcohol was prepared in a HLADH/G6PDH system, in which G6P [166] or G6S [165] was used as the regenerating cosubstrate. NAD(P)H was more stable with G6S than with G6P [165]. In these examples, optically inactive isotope-labeled compounds, as substrates in the main reaction, were converted to optically active isotope-labeled compounds.

Glutamic-α-d_1 acid was synthesized from α-ketoglutarate and NH_4^+ either in a glutamate dehydrogenase (GlutDH)/aldehyde dehydrogenase or GlutDH/YADH system with NADH cycling [169]. Glycoaldehyde-1,2,2-d_3 and ethanol-1,1-d_2 were regenerating cosubstrates, respectively. (R)-Trifluoroethanol-1-d_1 was prepared in a HLADH/FDH system with DCO_2Na as a cosubstrate [169]. In these examples, isotopes came from cosubstrates for the coenzyme regeneration.

2.3.1.5 Pharmaceuticals and Others

VANDECASTEELE [23] used carnitine dehydrogenase to convert 3-dehydrocarnitine to L-carnitine, a compound which is of nutritional and pharmacological interest. For NADH regeneration a chemical method with sodium dithionite, enzymatic methods with YADH, YADH plus aldehyde dehydrogenase, and GDH were compared. GDH gave the best result both in the cycling number (530) and in the cycling rate of NADH (7.4 h^{-1}).

12-Ketochenodeoxycholic acid, an intermediate in the synthesis of chenodeoxycholic acid utilized for dissolution of cholesterol gall-stones, can be en-

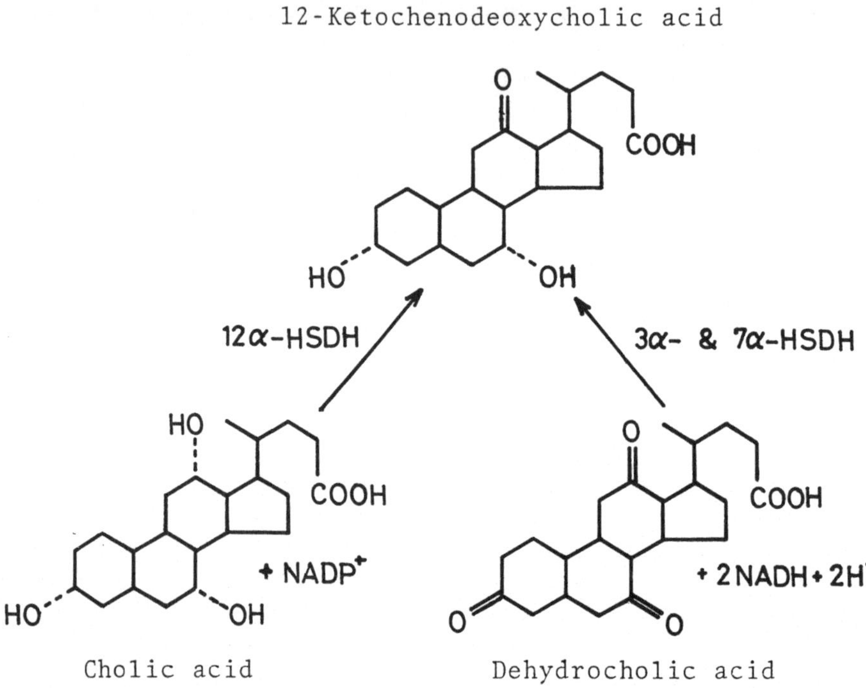

12-Ketochenodeoxycholic acid

Fig. 2.7. Production of 12-ketochenodeoxycholic acid (Abbreviation: HSDH, hydroxysteroid dehydrogenase)

zymatically synthesized either by the reduction of dehydrocholic acid or by the oxidation of cholic acid (Fig. 2.7). The former approach was attempted using 3α- and 7α-hydroxysteroid dehydrogenase with NADH regeneration by FDH [154]. Complete transformation of dehydrocholic acid to 12-ketochenodeoxycholic acid was achieved with an NADH cycling number up to 1200.

A soluble methane monooxygenase from methane-grown *Methylosinus sp.* CRL 31 was used for the production of epoxides from alkenes. For regeneration of NADH, FDH, diol dehydrogenase, primary ADH, and secondary ADH were compared [150, 151]. Almost the same results were obtained for all of these systems for NADH regeneration. The biochemical transformation of alkenes to epoxides, especially propylene to propyleneoxide, is of great industrial interest [148, 149].

Ethyl-(*R*)-4-chloro-3-hydroxybutanoate, (*R*)-2,2,2-trifluoro-L-phenylethanol, ethyl-(*S*)-3-hydroxyvalerate, (*S*)-lactoaldehyde dimethyl acetal, and (*S*)-3-hydroxybutanol dimethyl acetal were synthesized by using HLADH, YADH, or NADP-dependent ADH from *Thermoanaerobium brockii* [170]. NAD(P)H regeneration was carried out by GDH. The NAD(P)H cycling numbers were from 72 to 90.

Threo-D_S(+)-Isocitrate may be a useful starting material in chiral synthesis, and was produced from α-ketoglutarate and carbon dioxide by using isocitrate dehydrogenase and G6PDH with G6P [166] or G6S [165] as cosubstrates for

114

NAD(P)H regeneration. The NAD(P)H regeneration system could be replaced by a conjugated system of FDH/diaphorase, FDH/ADH, or FDH/glutamate dehydrogenase. These systems involve a transhydrogenation reaction between NADH and NADPH. The cycling numbers of coenzymes were between 1000 and 1500.

There exists an interesting versatile enzyme called enorate reductase [178–185], which catalyzes the following reactions:

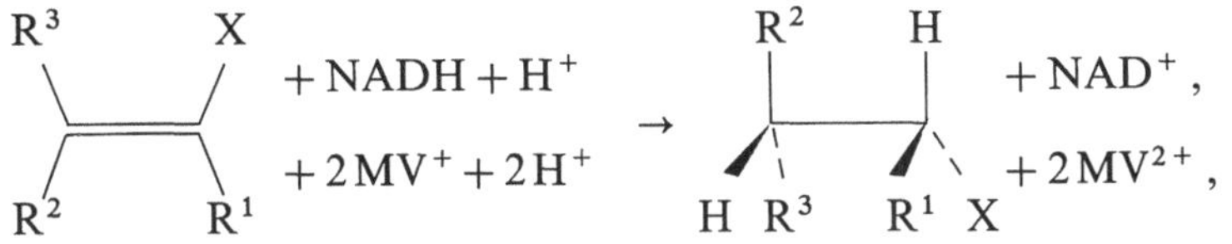

$$\begin{array}{ccc} & & + \, NADH + H^+ \\ & \rightarrow & \\ & & + \, 2\,MV^+ + 2\,H^+ \end{array} \qquad \begin{array}{c} + \, NAD^+, \\ \\ + \, 2\,MV^{2+}, \end{array} \qquad \begin{array}{c} (2.4) \\ \\ (2.5) \end{array}$$

$$NAD^+ + 2\,MV^+ + H^+ \rightarrow NADH + 2\,MV^{2+}. \tag{2.6}$$

X is either COO^- or CHO. As for R^1, R^2, and R^3, various functional groups are allowed. Therefore, this enzyme has a wide applicability for chiral reductive synthesis. The reaction in Eq. (2.5) shows the possibility of the direct coupling of this enzyme reaction with the electrode regeneration reaction of methyl viologen.

2.3.2 Process with Regeneration of NAD(P)$^+$

Enzyme processes with regeneration of NAD(P)$^+$ studied so far are listed in Table 2.9. Various types of 3S lactones were obtained from the corresponding pentane-1,5-diols with a prochiral center at C-3. The reaction was catalyzed by HLADH [187]. Figure 2.8 shows the transformation of 3-methylpentane-1,5-diol to (3S)-3-methylvalerolactone. The oxidized pyridine coenzyme, NAD$^+$, was regenerated chemically by FMN. Using the same system, oxidations of monocyclic meso diols to enantiometrically pure chiral lactones were carried out [188–190]. Bicyclic lactones were also prepared using the same method [191].

Steroid transformation is a area of practical interest for the application of NAD(P)$^+$ cycling. Testosterone was converted to 4-androstene-3,17-dione by a β-hydroxysteroid dehydrogenase/LDH system with pyruvate as a cosubstrate for NAD regeneration. This reaction was carried out, in a two-phase system of water and butylacetate, which increased the solubility of the substrate (testosterone) [192]. The NAD cycling number was 44 in this case. Androsterone was converted to 5α-androstane-3,17-dione by α-steroid dehydrogenase conjugated either with a chemical method of NAD regeneration by phenazine methosulfate [193] or with

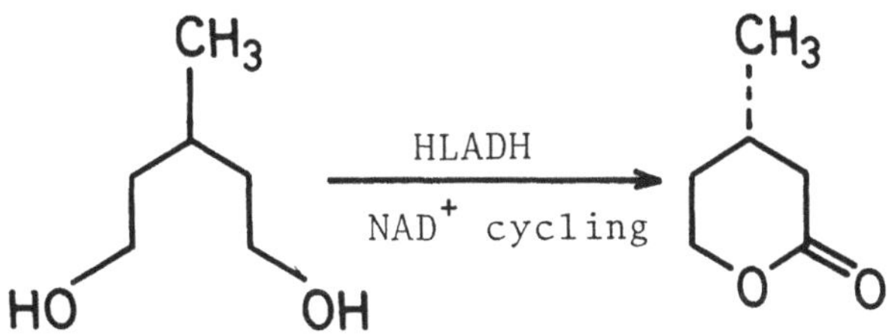

Fig. 2.8. Transformation of 3-methylpentane-1,5-diol to (3S)-3-methylvalerolactone by horse liver alcohol dehydrogenase

115

Table 2.9. Enzyme processes with regeneration of $NAD(P)^+$

Main product	Enzyme for main reaction	Method for coenzyme regeneration	Main substrate	Cosubstrate	NAD cycling Number [–]	NAD cycling Rate [h^{-1}]	Productivity [mol/l/h]	Remark[a]	Ref.
3S Lactones	HLADH	FMN	Pentane-1,5-diols	O_2	< 10	< 0.43	–	B	[187]
Chiral γ-lactones	HLADH	FMN	Monocyclic meso diols	O_2	< 12.5	< 0.26	–	B	[188–190]
Bicyclic lactones	HLADH	FMN	Diols	O_2	< 20	< 0.3	–	B	[191]
4-Androstene-3,17-dione	β-Hydroxy-steroidDH	LDH	Testosterone	Pyruvate	44	0.96	–	B	[192]
5α-Androstane-3,17-dione	α-SteroidDH	PMS	Androsterone	O_2	0.59	0.59	–	B, IMC, IME	[193]
12-Ketocheno-deoxycholic acid	12α-hydroxy-steroidDH	GlutamateDH	Cholic acid	α-Keto-glutarate, NH_3	1600	16.7	–	B, IME, NAD or NADP	[194]
Geranial	HLADH	PMS	Geraniol	O_2	84	0.7	–	B, IMC, IME	[195]
Geranial	HLADH	LDH	Geraniol	Pyruvate	400	4	–	B, IMC, IME	[195]
Geranial	HLADH	HLADH	Geraniol	Acetaldehyde	1300	7.6	–	B, IMC, IME	[195]
Fructose	SorbitolDH	YADH	Sorbitol	Acetaldehyde	10000	2860	2.9E–3	C, IME	[196]
5α-Androstane-3,17-dione	2α-Hydroxy-steroidDH	NADH-oxidizing bacterium	Androsterone	O_2	7.6	0.76	–	B, IME	[197]
Pyruvate	L-LDH	Pretreated vitreous carbon	Lactate	$-e^-$	18	1.1	–	B, IME on electrode	[198]
Acetaldehyde	YADH	Pt gauze	Ethanol	$-e^-$	2.57	0.86	–	B, IME	[199, 200]
Acetaldehyde	YADH	Diaphorase	Ethanol	Ferricyanide	0.38	0.45	–	B, IMC, IME	[201]
Acetaldehyde	YADH	Respiratory (E. coli) NADH oxidase	Ethanol	O_2	100	–	–	B, IME	[202]
G6P lactone	G6PDH	Respiratory (E. coli) NADH oxidase	G6P	O_2	13	7.4	–	B, IME, NADP	[203]

[a] Abbreviations: B, batchwise operation; C, continuous operation; IMC, immobilized coenzyme; IME, immobilized enzyme.

an NADH-oxidizing system in bacteria [197]. The ultimate electron acceptor was molecular oxygen in both cases. NAD cycling numbers obtained, however, were as low as 0.59 and 7.6, respectively. As was shown before, 12-ketochenodeoxy cholic acid could be synthesized by the reduction of dehydrocholic acid with NADH cycling [154]. Alternatively, this compound could be synthesized from the oxidation of cholic acid by 12α-hydroxysteroid dehydrogenase with $NADP^+$ cycling (Fig. 2.7). Glutamate dehydrogenase with ketoglutarate and ammonia was used as a regeneration system for the oxidized coenzyme [194]. A moderate NADP cycling number of 1600 was obtained. NADP, however, is more expensive than NAD by a factor of about ten.

Geranial, a flavoring compound, was formed from geraniol by using HLADH [195]. A chemical method by phenazine methosulfate, a coupled enzyme method with LDH/pyruvate, and a substrate conjugation method with acetaldehyde were compared as regeneration systems of NAD. As expected, the substrate conjugation method gave the best result in the NAD cycling number (1300) and the chemical method the poorest (84).

Fructose was prepared continuously from sorbitol by using a sorbitol dehydrogenase/YADH system with acetaldehyde as a cosubstrate [196]. The two enzymes were immobilized on Sepharose, which was packed in a column. To this, the substrate solution containing NAD at a low concentration level was fed continuously. The NAD cycling number obtained was as high as 10000 with 99% conversion of the substrate, sorbitol.

COUGHLIN et al. [199, 200] coupled the YADH reaction with a direct electrochemical regeneration of NAD^+ on a platinum gauge electrode. Acetaldehyde was produced from ethanol with an NAD cycling number of 2.6. LAVAL et al. [198] also attempted a direct electrochemical regeneration of NAD^+ on pretreated vitreous carbon electrode. This electrochemical process was conjugated with LDH reaction to produce lactate from pyruvate. The NAD cycling number was 18. These approaches for the direct electrochemical NAD^+ regeneration seem to have a problem in the selectivity of the electrochemical oxidation of NADH to NAD^+ [74]. This may partly explain the relatively low NAD cycling number in these methods.

MAY et al. [201] encapsulated NAD together with YADH and diaphorase in hydrocarbon-based liquid surfactant membrane microcapsules, which could retain native NAD without leakage. When ethanol and ferrocyanide were added externally, this system produced acetaldehyde. However, the NAD cycling number was as low as 0.38, probably because of the poor permeability of the substrate through the liquid membrane. The NADH-oxidizing activity of *Escherichia coli* linked with the respiratory system was coupled either with YADH [202] or G6PDH [203]. NAD^+ was cycled in the former case and $NADP^+$ in the latter case. The cycling numbers of coenzymes were 100 and 13, respectively.

2.3.3 Process with Regeneration of ATP

Enzyme processes with regeneration of ATP are listed in Table 2.10. CAMPBELL et al. [204] coimmobilized hexokinase and pyruvate kinase within semipermeable

Table 2.10. Enzyme processes with regeneration of ATP

Main product	Enzyme for main reaction	Method for coenzyme regeneration	Main substrate	Cosubstrate	ATP cycling Number [–]	ATP cycling Rate [h⁻¹]	Productivity^a [mol/l/h]	Remarks	Ref.
Pyruvate	Pyruvate kinase	Hexokinase	Phosphoenol-pyruvate	Glucose	3	2	$6.4E-5$	B, IME	[204]
Glucose-1-phosphate	Galactokinase	Pyruvate kinase	Galactose	Phospho-enol-pyruvate	–	–	–	B, IME	[205]
Glutathione	Glutathione synthesis system (S. cerevisiae)	Glycolytic pathway (S. cerevisiae)	L-Glutamate, L-Cysteine, Glycine	Glucose	–	–	–	C, IME	[206]
Glutathione	Glutathione synthesis system (E. coli)	Acetate kinase (E. coli)	L-Glutamate, L-Cysteine, Glycine	Acetyl-phosphate	<0.497	<0.002	–	C, IMC, IME	[207]
Glutathione	Glutathione synthesis system (E. coli)	Acetate kinase (E. coli)	L-Glutamate, L-Cysteine, Glycine	Acetyl-phosphate	~0.11	0.011	–	C, IME	[208]
Glutathione	Glutathione synthesis system (E. coli)	Acetate kinase (E. coli)	L-Glutamate, L-Cysteine, Glycine	Acetyl-phosphate	~2.98	0.0155	–	C, IMC, IME	[209]
Glutathione	Glutathione synthesis system (S. cerevisiae)	Glycolytic pathway (S. cerevisiae)	L-Glutamate, L-Cysteine, Glycine	Glucose	–	–	–	C, IMC, IME	[210]
NADP	NAD kinase (Brevibacterium ammoniagenes)	Glycolytic pathway (S. cerevisiae)	NAD	Glucose	1~3	~0.9	–	C, IME	[212]
NADP	NAD kinase (chicken liver)	Acetate kinase (E. coli)	NAD	Acetyl-phosphate	8.35	2.28	$2.28E-4$	C, IME	[213]
NADP	NAD kinase	Acetate kinase	NAD	Acetyl-phosphate	~285	~0.74	–	B, IME	[214]
CDP choline	Choline kinase, CDP-choline,	Glycolytic pathway	CMP, Choline chloride	Glucose	–	–	–	B, IME	[215]

	pyrophosphorylase (Yeast)	(Yeast)						
CDP choline	Choline kinase, CDP-choline, pyrophosphorylase (Yeast)	Glycolytic pathway (Yeast)	CMP Choline chloride	Glucose	–	–	–	B, IMC, IME Need for NAD [216]
CDP choline	Choline kinase, CDP-choline, pyrophosphorylase (Yeast)	Glycolytic pathway (Yeast)	CMP Choline chloride	Glucose	7000	1400	6.0E−3	C, IMC, IME [217]
CDP choline	Choline kinase, CDP-choline pyrophosphorylase (Hydrocarbon-assimilating yeast)	Glycolytic pathway	Phosphoryl choline	Glucose	–	–	–	B, Whole cell (lyophilized) [219, 220]
Creatine phosphate	Creatine kinase	Acetate kinase	Creatine hydrate	Diammonium acetyl-phosphate	~46.8	~1.3	1.66E−3	B, IME [221]
Glucose-6-phosphate	Hexokinase	Acetate kinase	Glucose	Diammonium acetyl-phosphate	~109	~2.18	7.0E−3	B, IME [222]
L-Glycerol-3-phosphate	Glycerol kinase	Acetate kinase	Glycerol	Diammonium acetyl-phosphate	~53	~0.396	7.0E−4	B, IME [223]
L-Glycerol-3-phosphate	Glycerol kinase	Carbamate kinase	Glycerol	Carbamoyl-phosphate	~3.6	–	–	C, IME [224]
Glucose-6-phosphate	Hexokinase	Acetate kinase	Glucose	Acetyl-phosphate	12000	~28	–	C, IMC, IME [225]

[a] Abbreviations: B, batchwise operation; C, continuous operation; IMC, immobilized coenzyme; IME, immobilized enzyme.

119

collodion microcapsules. ATP could be cycled in this system three times in 90 min. CHANG et al. [205] attempted galactose removal with ATP cycling in a galactose kinase/pyruvate kinase system immobilized in microcapsules. This system may be applicable for the treatment of galactosemia.

Glutathione, a physiologically and medically important tripeptide, is biosynthesized by glutathione synthetase. MURATA et al. [206, 210] combined the glutathione synthesizing system in *Saccharomyces cerevisiae* with the glycolytic pathway in the same organism as a regenerator of ATP. Intact cells of *S. cerevisiae* were entrapped in a polyacrylamide gel. A column containing the gel with the immobilized cells could produce glutathione continuously with endogeneous ATP cycling when the constitutional amino acids, L-glutamate, L-cysteine, and glycine, were supplied with glucose as an energy source. There was no need for an external feed of ATP. The endogeneous ATP seemed to remain within the immobilized cells.

Immobilized *Escherichia coli* containing a glutathione synthesizing system and acetate kinase were also used for the same purpose [207–209]. ATP was either supplied externally or coimmobilized in a dextran-bound form together with enzymes in polyacrylamide gel. The energy source, acetylphosphate, was supplied externally. Glutathione was produced continuously but the ATP cycling number was not very high.

NADP is an expensive reagent and could be synthesized enzymatically from NAD catalyzed by NAD kinase. MURATA et al. [212] used NAD kinase in *Brevibacterium ammoniagenes* together with the glycolytic pathway of *Saccharomyces cerevisiae*. These systems altogether were coimmobilized in a polyacrylamide gel, which produced NADP continuously when AMP was fed externally. MIYAWAKI et al. [213] coimmobilized NAD kinase and acetyl kinase in an ultrafiltration hollow fiber tube. Upon the continuous feed of a solution containing NAD, acetylphosphate, and ATP at a low concentration level, NADP was produced with an ATP cycling number of 8.4. WALT et al. [214] employed the same enzyme system for the same purpose. Both enzymes were immobilized in poly(acrylamide-co-N-(acryloxy)succinimide) gel. NADP was produced in a batch operation with an ATP cycling number of 285.

CDP-choline is known as a precursor of a phospholipid, lectithin, and is used as a therapeutic agent for brain nerve injury [217]. Some yeast cells contain cholin kinase and CDP-choline pyrophosphorylase along with the glycolytic pathway. These enzymes can be used for the production of CDP-choline in a way as shown in Fig. 2.9. KIMURA et al. [215] immobilized dried yeast cells containing these enzymes in a photo-crosslinked gel of polyethylene glycol hydroxyethylacrylate. The immobilized cells could produce CDP-choline with the addition of CMP and choline chloride together with ATP and NAD. ADO et al. [217] entrapped *Saccharomyces cerevisiae* cells in microcapsules made of cellulose acetate butyrate and chitosan. With the feed of a solution containing glucose, CMP, and choline chloride, a column packed with the microcapsules immobilizing the *Saccharomyces* cells produced CDP-choline continuously with no need for external addition of ATP and NAD. The endogeneous ATP was efficiently cycled and the ATP cycling number estimated was 7000 in 5 hrs. The CDP-choline-producing activity of hydrocarbon-assimilating yeast was also investigated [219, 220].

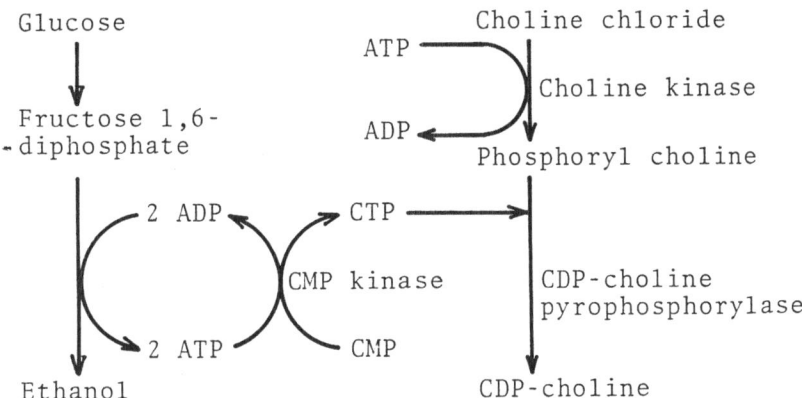

Fig. 2.9. Synthesis of CDP-choline by *Saccharomyces cerevisiae* involving ATP cycling

Chemical compounds with high energy phosphate-bonds could be produced with ATP cycling. Creatine phosphate was prepared in a conjugated system of creatine kinase and acetate kinase, both immobilized in a polyacrylamide gel [221]. With the addition of creatine hydrate, acetylphosphate, and a small amount of ATP, creatine phosphate was produced with an ATP cycling number of 47. Immobilized systems of hexokinase/acetylkinase and glycerol kinase/acetylkinase were also used to produce G6P [222] and L-glycerol-3-phosphate [223], respectively. Cycling numbers of ATP were 109 and 53 in these cases. Carbamate kinase was also used as a regenerator of ATP to produce L-glycerol-3-phosphate by glycerol kinase [224]. BERKE et al. [225] used polyethyleneglycol-bound ATP, which was coimmobilized together with hexokinase and acetylkinase in a membrane reactor. G6P was continuously produced for 18 days with an ATP cycling number of 12000. However, the enzymes should be added periodically to counteract the inactivation of enzymes.

Except for the case with glutathione, the investigations exemplified here are all relevant to the phosphate transfer from ATP by kinases. Synthetases are another important enzyme group involving ATP. The enzymes in this group function as catalysts in the synthetic reaction by use of the high-energy potential of nucleotide triphosphates. In connection with ATP cycling, we refer to gramicidin S (Fig. 2.10) catalytically formed by gramicidin synthetase [226] and enniatin (Fig. 2.11) formed by enniatin synthetase [227].

Another investigation of interest involving ATP is cell-free protein synthesis by posterior silk gland polyribosome [229, 230]. These polyribosomes were immobilized on DEAE-Sephadex to form a genetic information-transducing bioreactor, which successfully could produce silk fibroin from corresponding amino acids and ATP.

Investigations involving the cycling of both NAD(P) and ATP also exist. WONG et al. [240] enzymatically prepared ribulose 1,5-biphosphate, an expensive substrate for ribulose-biphosphate carboxylase, from 6-phosphogluconate (6-PG) as shown in Fig. 2.12. This method starting from 6-PG is superior to that

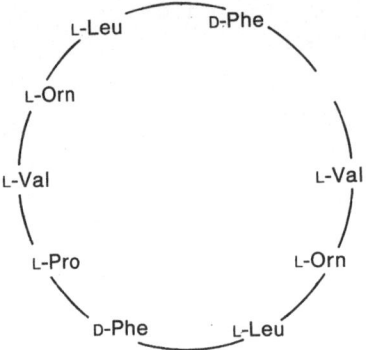

Fig. 2.10. Structure of Gramicidin S

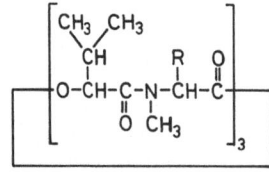

Fig. 2.11. Structure of enniatin homologues. (Enniatin A, $R=CH(CH_3)CH_2CH_3$; Enniatin B, $R=CH(CH_3)_2$; Enniatin C, $R=CH_2CH(CH_3)_2$)

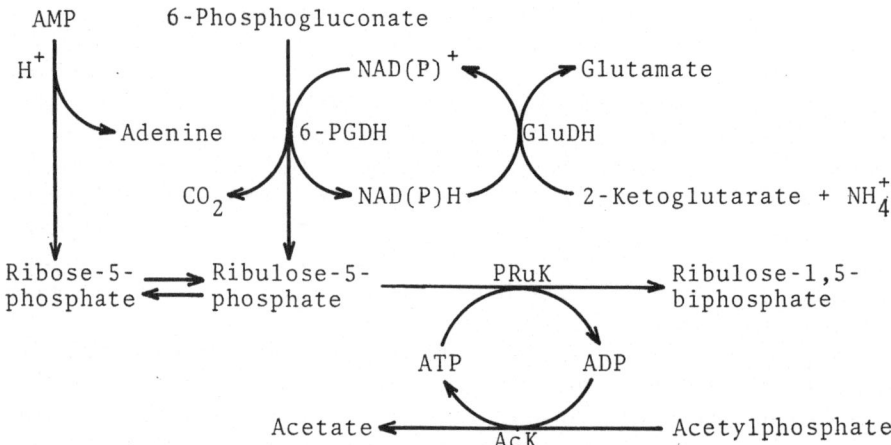

Fig. 2.12. Synthesis of ribulose-1,5-biphosphate with cycling of ATP and NAD(P). (Abbreviations: 6-PGDH, 6-phosphogluconate dehydrogenase; GluDH, glutamate dehydrogenase; PRuK, phosphoribulokinase; AcK, acetate kinase)

starting from AMP (also shown in Fig. 2.12) when 6-PG is available for some other reason. ABRIL et al. [241] studied an enzymatic process for the synthesis of adenosine 5'-*O*-(3-thiotriphosphate) (ATP-γ-S), an ATP analog useful in mechanistic enzymology. This process started from dihydroxyacetone, sodium thiophosphate, ADP, and phosphoenol pyruvate. The starting material, dihydroxyacetone, was inexpensively available. The yield of ATP-γ-S based on ADP was much improved (80%) by recycling the by-product, ATP.

2.3.4 Coenzyme Cycling in Analytical Use

2.3.4.1 Coenzyme Cycling as a Device for Chemical Amplification

Coenzyme cycling as a device for chemical amplification, as listed in Table 2.11, was first attempted by LOWRY et al. [294] for a highly sensitive method of micro-assay for NAD and NADP. In this technique, a signal (a concentration of a substance to be detected) is chemically amplified in a cycling reaction catalyzed by conjugated enzymes to produce an indicator (a final product of the cycling reaction). Thus the original signal is amplified as a concentration of the indicator. The indicator should be a substance which is easily detectable in an indicator reaction. This technique is well exemplified in a cycling assay of $NADP^+$ and/or NADPH [294]. In this case, the following two reactions occurred in the cycling reaction in which a small amount of $NADP^+$ and/or NADPH, the substance to be detected, produced a large amount of 6-phosphogluconate, the indicator.

$$NADPH + \alpha\text{-ketoglutarate} + NH_4^+ \rightarrow NADP^+ + glutamate, \qquad (2.7)$$

$$NADP^+ + glucose\text{-}6\text{-phosphate} \rightarrow NADPH + 6\text{-phosphogluconate}. \quad (2.8)$$

These two reactions were catalyzed by glutamate dehydrogenase and G6PDH, respectively. After a period of 30 min of coenzyme cycling, the reaction mixture was exposed to a high temperature (100 °C) to stop the cycling reaction. 6-Phosphogluconate, thus produced, was analyzed in the following indicator reaction:

$$6\text{-phosphogluconate} + NADP^+ \rightarrow D\text{-ribulose 5-phosphate}$$
$$+ CO_2 + NADPH. \qquad (2.9)$$

This reaction, catalyzed by 6-phosphogluconate dehydrogenase, was carried out with an excess amount of $NADP^+$. The NADPH produced was analyzed fluoro-metrically.

In this case, each molecule of NADP catalyzed the formation of 5000 to 10 000 molecules of 6-phosphogluconate in 30 min. Thus NADP at a quantity of 10^{-13} mol could be detected with an amplification factor or coenzyme cycling number of 5000 to 10 000. In addition, NADPH produced in the indicator reaction might be cycled again in a double cycling step. By using this technique, the amplification factor went up as high as 5×10^7 so that NADP at a level of 10^{-16} mol could be detected. In the same manner, a triple cycling is also possible. This may provide a microassay technique in which the existence of even a single molecule of coenzyme is detectable, in principle [298]. Therefore, the coenzyme cycling technique can be considered as one of the most sensitive microassays available at present. The detection limits exemplified here may be compared with those in fluorometry (10^{-10} mol) and isotope radiometry (10^{-13} mol) [298].

This technique of coenzyme cycling was applied for microassays of NAD^+, NADH, $NADP^+$, and NADPH in rat liver, brain, and blood [294]. The reduced and the oxidized state of pyridine nucleotides were distinguishable from each

Table 2.11. Coenzyme cycling for analytical use: cycling as a device for chemical amplification

Substance to be detected	Substance to be cycled	System for cycling	Indicator	Detection limit [mol/l]	Detection limit [mol]	Ref.
NADP	a	G6PDH/GlutamateDH	6-P-Gluconate	6E−9	6E−13	[294]
NAD	a	LDH/GlutamateDH	Pyruvate	2E−9	2E−13	[294]
NAD	a	ADH/MDH	Malate		1E−12	[295]
CoA	a	Phosphotransacetylase/citrate synthase	Citrate		4E−14	[297]
NAD, NADP	a	ADH/ADH	Acetaldehyde		2E−12	[299]
NAD	a	G6PDH/NAD-peroxidase	6-P-Gluconate		3E−14	[300]
NAD	a	ADH–DIA conjugate	Reduced dichlorophenol-indophenol		3E−12	[302]
Glucose/gluconolactone	a	Glucose oxidase/GDH	Oxygen	8E−7	2E−9	[309]
L-Lactate/pyruvate	a	Cytochrome b_2/LDH	$[Fe(CN)_6]^{4-}$	3E−7	6E−10	[309]
L-Lactate/pyruvate	a	Lactate oxidase/LDH	Oxygen	8E−8	8E−10	[317]
Bile acid	NAD	3β- or 17β-Hydroxy-steroidDH	Testosterone	2E−6	3E−13	[312]
3α- and 3β-Hydroxy-steroid/3-ketosteroid	NAD	ADH/MDH	Malate		2–4E−13	[314]
Gentamycin	Gentamycin-N⁶-(2-carboxyethyl)NAD	G6PDH/NAD-peroxidase	6-P-Gluconate		5E−12	[300]
Biotin	Biotinyl-aminoethyl NAD	LDH/DIA	Reduced thiazonyl blue	1E−10		[311]
2,4-Dinitrophenyl 6-aminocaproate	2,4-Dinitrophenyl-aminoethyl NAD	LDH/DIA	Reduced thiazonyl blue	1E−10		[311]
Human serum albumin	Fructose-6-phosphate	Phosphofructokinase/fructose-1,6-biphosphatase	ADP			[313]

a Substance to be cycled is the same as the substance to be detected.

other by using selective acid-decomposition of NAD(P)H or alkali-decomposition of $NAD(P)^+$.

KATO et al. [295] employed MDH/ADH as a cycling system in an NAD assay. SCHULMAN et al. [299] used a substrate conjugation technique as a cycling system of NAD. A single enzyme, ADH, catalyzed both reduction and oxidation of NAD by using conjugated substrates of ethanol and lactoaldehyde. This made the system simple. Cox et al. [300] used NAD^+-peroxidase in the oxidation cycle of NADH. CREMONESI et al. [302] prepared an ADH-DIA-albumin conjugate for the effective cycling of NAD.

The same cycling technique was also applied for an assay of CoASH and/or acetyl CoA at a level of 4×10^{-14} mol [297]. In this case, the following cycling reactions were employed:

$$CoASH + acetylphosphate \rightarrow acetyl\ CoA + phosphate, \qquad (2.10)$$

$$acetyl\ CoA + oxaloacetate \rightarrow CoASH + citrate. \qquad (2.11)$$

The former reaction was catalyzed by phosphotransacetylase and the latter by citrate synthetase. Citrate, as a detector, was fluorometrically detected in the following reactions catalyzed by aconitase and isocitrate dehydrogenase, respectively:

$$citrate \rightarrow isocitrate, \qquad (2.12)$$

$$isocitrate + NADP^+ \rightarrow \alpha\text{-ketoglutarate} + CO_2 + NADPH + H^+. \qquad (2.13)$$

CoASH and acetyl CoA in the original sample were separately detectable by using either N-ethylmaleimide as a scavenger of CoASH or Sepharose (glutathione-2-pyridyl disulfide) conjugate as a scavenger of acetyl CoA.

The cycling technique described above is not restricted only to coenzyme. Anything other than coenzyme may be cycled and chemically amplified if an appropriate cycling reaction is available. Chemical amplifications of glucose/gluconolactone and L-lactate/pyruvate were studied using a glucose oxidase/GDH system and cytochrome b_2/lactate dehydrogenase system, respectively [309]. A system with lactate oxidase/LDH was also used for the chemical amplification of L-lactate/pyruvate [317]. In these cases, electrode reactions were used for the indicator reaction.

Because of the high sensitivity of the coenzyme cycling technique, this method may be applied in the immunoassay with coenzyme label. For this purpose, gentamycin-N^6-(2-carboxyethyl) NAD^+ was prepared. This compound could be cycled and amplified in an enzyme system with G6PDH/NAD^+-peroxidase. The cycling rate was inhibited by anti-gentamycin antibody and the inhibition was reversed in competitive binding reactions with free gentamycin. Thus gentamycin at a level of 5.2×10^{-12} mol was detected [300]. The same type technique was used for the assay of biotin and 2,4-dinitrophenyl-6-aminocaproate by using systems with biotinyl-N^6-(2-aminoethylamino)-NAD^+/avidin (= biotin-binding protein) and 2,4-dinitrophenyl-N^6-(2-aminoethylamino)-NAD^+/antibody to 2,4-dinitrophenyl residue [311], respectively.

As the coenzyme in a different state (e.g. oxidized or reduced) in the cycling reaction is separable, the coenzyme cycling technique is also applicable for the assay of a substrate for an enzyme reaction in which a coenzyme at either state of the two in the cycling reaction is produced stoichiometrically from the reaction with the substrate. Bile acid was analyzed by this method [312]. Bile acid, first, was completely oxidized with an excess amount of NAD^+ catalyzed by 3α-hydroxysteroid dehydrogenase. Then the residual NAD^+ was decomposed by alkalinization and heat. Thus bile acid (original signal) was totally converted to the equal molecular amount of NADH (intermediate signal), which was cycled in a coupled oxidation-reduction of dehydroepiandrosterone catalyzed by a single enzyme, 3β- or 17β-hydroxysteroid dehydrogenase. The final product, $\Delta 5$-testosterone, as the indicator was enzymatically analyzed. The sensitivity limit for this method was 3×10^{-13} mol. The same type technique was also used for the measurement of 3α- and 3β-hydroxysteroids, as well as 3-ketosteroid at a level of $2-4 \times 10^{-13}$ mol [314].

The coenzyme cycling technique can be also applied to the activity assay of enzymes which catalyze the production of coenzyme at either of the two states in a cycling reaction. KATO and LOWRY [296] used this technique for the activity measurement of enzymes such as hexokinase, phosphofructokinase, G6PDH, 6-phosphogluconate dehydrogenase, NADP-linked isocitrate dehydrogenase, MDH, LDH, glutamate dehydrogenase, and ATP:NMN adenyltransferase in nuclei, cytoplasm, and cell bodies of single dorsal root ganglian cells of the rabbit.

2.3.4.2 Coenzyme Cycling for the Reduction of the Necessary Amount of Coenzyme

The other purpose of coenzyme cycling in analytical usage is the reduction of the necessary amount of expensive coenzyme, in order to reduce the cost of enzyme assays (Table 2.12). For this purpose, a combination of chemically modified coenzymes with the electrochemical device as a detector is most frequently used. DAVIES and MOSBACH [143] prepared dextran-bound NAD, which was incorporated into an enzyme electrode system with glutamate dehydrogenase and LDH. As the electrode used was sensitive to the product (NH_4^+) from the coenzyme cycling reaction, this system could work as a glutamate or a pyruvate sensor with detection limits of 10^{-4} M and 2×10^{-5} M, respectively.

YAMAZAKI and MAEDA [235] prepared NAD-containing polymer, in which conjugated enzymes of ADH and DIA were entrapped. This polymer functioned as an ethanol sensor with a detection limit of 1.2×10^{-6} mol.

PAU and RECHNITZ [307] incorporated dextran-bound NAD into an L-alanine dehydrogenase/L-LDH system immobilized at the surface of a potentiometric ammonia gas sensor. This system functioned as an L-alanine sensor with a detection limit of 10^{-7} mol and a lifetime of at least 10 days.

HUCK et al. [304] prepared a 3-β-naphthoyl-Nile Blue-modified graphite electrode which was capable of catalyzing the electrochemical oxidation of NADH to NAD^+. With ADH and NAD, this electrode could be used as an ethanol sen-

Table 2.12. Coenzyme cycling in analytical use; for reduction of the necessary amount of coenzyme

Substance to be detected	Substance to be cycled	System for cycling	Indicator	Detection limit [mol/l]	Detection limit [mol]	Ref.
Ethanol	NAD	ADH/3-β-naphthoyl–Nile Blue modified electrode	e^-	1E–5		[304]
Alanine	NAD	AlanineDH/3-β-naphthoyl–Nile Blue modified electrode	e^-	1E–6		[304]
Lactate	NAD	LDH/3-β-naphthoyl–Nile Blue modified electrode	e^-	1E–6		[304]
Glutamate	NAD	GlutamateDH/3-β-naphthoyl–Nile Blue modified electrode	e^-	1E–6		[304]
Leucine	NAD	LeucineDH/3-β-naphthyl–Nile Blue modified electrode	e^-	1E–6		[304]
Glutamate	Dextran–NAD	LDH/GlutamateDH	NH_3	1E–4	5E–6	[143]
Pyruvate	Dextran–NAD	LDH/GlutamateDH	NH_3	2E–5	1E–6	[143]
Ethanol	Matrix–NAD	ADH/DIA	Resorufin	1E–2	1E–6	[235]
L-Alanine	Dextran–NAD	L-AlanineDH/LDH	NH_3	1E–5	1E–7	[307]

sor. In this case, the chemically modified electrode with 3-β-naphthoyl-Nile Blue worked both as a regenerator of coenzyme and as a detector. In the same manner, alanine, lactate, glutamate, and L-leucine at a level of 10^{-6} M were also analyzed with the same electrode combined with alanine dehydrogenase, LDH, glutamate dehydrogenase, and L-leucine dehydrogenase, respectively.

YAO and MUSHA [310] immobilized NAD^+ at the surface of a carbon paste electrode. The bound NAD^+ could be reduced to NADH by ethanol (catalyzed by ADH) or by L-lactate (catalyzed by L-LDH). The NADH formed at the electrode surface was reoxidized electrochemically and the corresponding peak area in a linear-sweep voltammetry was proportional to the original amount of ethanol or L-lactate. In this case, the detection limit was as low as 5×10^{-11} mol although NAD immobilized was not cycled in the analytical process.

SAKAGUCHI et al. [306] prepared Dextran-bound NAD which was used for flow assays of lactate, glutamate, and glutamate-oxaloacetate aminotransferase in serum samples. The Dextran-NADH formed in the assay was recovered along with unreacted Dextran-NAD^+, reoxidized by phenazine methosulfate, and reused again. In this case, the expensive coenzyme was not cycled in situ but it was cycled in the whole assay system.

2.4 Chemically Modified Coenzymes

2.4.1 Chemically Modified NAD

Before discussing the bioreactor system with coenzyme cycling, chemically modified coenzymes, especially water-soluble macromolecule-bound coenzymes, should be mentioned [242–250]. This technique is highly relevant for affinity chromatography, in which coenzymes are immobilized on solid support materials while the binding activity to apoenzymes is maintained. In the present case, however, the catalytic activity of the coenzyme along with the binding activity should be maintained throughout the chemical modification process. For this purpose, chemical modifications at 6-NH_2, C^6, and C^8 in the adenine ring (Table 2.1) are possible choices. This was suggested from structure analysis of the enzyme-coenzyme-substrate complex of LDH [251, 252]. This rule is believed to hold for all other adenine coenzymes such as NADP and ATP.

LARSSON and MOSBACH [257] combined NAD, via ε-amino caproic acid as a spacer, with Sepharose using dicyclohexyl carbodiimide. The Sepharose-bound

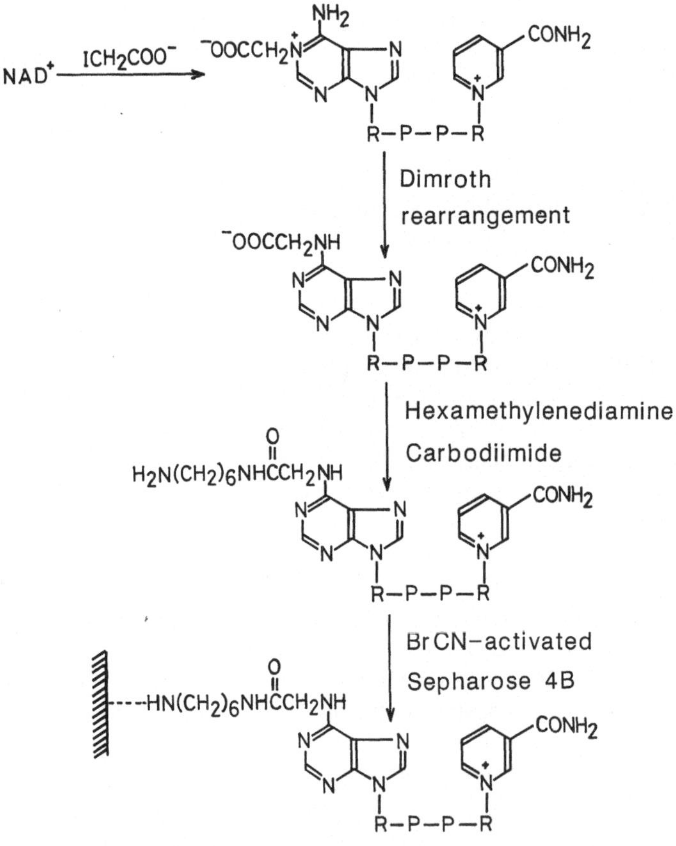

Fig. 2.13. Synthesis of Sepharose-bound NAD

Table 2.13. Macromolecule-bound NAD derivatives

Macromolecule	Spacer in adenine ring	Relative activity with			Ref.
		ADH	LDH	MDH	
Polyethyleneimine	$N^6HCOCH_2CH_2COOH$	0.97			[260]
Dextran T40	$N^6HCH_2CONH(CH_2)_6NH_2$	0.16	0.14		[259]
Sepharose 4B	$N^6HCH_2CONH(CH_2)_6NH_2$	0.007	0.001		[258, 259]
Polyethyleneimine	$N^6HCH_2CH(OH)CH_2COOH$	0.06	0.60		[261]
Polylysine	$N^6HCH_2CH(OH)CH_2COOH$	0.07	0.25		[261]
Polylysine	$N^6HCOCH_2CH_2COOH$	0.184			[29]
Polyethyleneimine	$C^8S(CH_2)_2COOH$	0.47	0.03		[262]
Polylysine	$C^8S(CH_2)_2COOH$	0.05	0.06		[262]
Dextran	$N^6H(CH_2)_2NH_2$	0.085	0.24	0.185	[263]
Sepharose	$N^6H(CH_2)_2NH_2$	–	–	–	[263]
Sepharose	$N^6H(CH_2)_2NHCO(CH_2)_5NH_2$	0.003	0.011	0.011	[263]
Affi Gel 10	$N^6H(CH_2)_2NH_2$	–	–	–	[263]
Aminopolyethylene-glycol	$N^6H(CH_2)_2COOH$	0.50	0.77	0.64	[265]
Dextran T70	$N^6H(CH_2)_2NH(COCH_2NH)_3H$	0.28	0.52	0.19	[266]
Methacrylcholine/(I) copolymer	3-[4-(2,3-Epoxypropoxy)butoxy]-2-hydroxypropyl acrylate (I) bound at N^6	0.37	0.045	0.26	[267]
N-Methacrylyl-2-glucosamine/(I) copolymer	3-[4-(2,3-Epoxypropoxy)butoxy]-2-hydroxypropyl acrylate bound at N^6		0.093	0.052	[267]
Polyethyleneglycol	$N^6H(CH_2)_2NH_2$				[268]
Bovine casein α_1	$N^6HCH_2CONH(CH_2)_6NH_2$	0.15			[269]
Bovine casein acetylated α_1	$N^6HCH_2CONH(CH_2)_6NH_2$	0.20			[269]
Bovine casein β	$N^6HCH_2CONH(CH_2)_6NH_2$	0.31			[269]
Dextran T10, 40, 70, 500	$N^6H(CH_2)_2COOH$	~1			[270]

coenzyme expressed only 0.2% of the coenzyme activity. The accurate mode of the binding of NAD to the matrix was unknown.

LINDBERG et al. [258] established a 4-step procedure for NAD derivation at 6-NH_2 in the adenine ring starting from (1) alkylation of NAD by iodoacetate to form N^1-carboxymethyl NAD$^+$, followed by (2) Dimroth rearrangement by alkali, (3) addition of a spacer (hexamethylenediamine) by carbodiimide reaction, and (4) binding to BrCN-activated Sepharose 4B as shown in Fig. 2.13. This method with the alkylation at the N^1 position in the adenine ring followed by DIMROTH rearrangement is widely used with modifications. LARSSON et al. [259] applied this method to combine NAD to Dextran T40 and Sepharose 4B. The coenzyme activity of Dextran-bound NAD was about 15% of that of native NAD with both YADH and LDH. However, the activity of Sepharose-bound NAD was less than 1%, reflecting the lower mobility of the bound NAD on the solid support. WYKES et al. [260] directly modified 6-NH_2 in the adenine ring by succinic anhydride to form succinyl-NAD, which was bound to polyethylene imine by using carbodiimide. This NAD derivative showed 97% coenzyme activity with

Table 2.14. NAD-bearing copolymers

NAD bearing monomer[a]	Monomer to be copolymerized	Relative activity with			Ref.
		ADH	LDH	MDH	
N1	Acrylamide (A)	0.18	<0.01	0.33	[271]
N1	Methacrylamide	0.52	0.65	0.74	[271]
N1	N-Acryloyl-6-aminohexanoic acid	0.18	0.02	0.11	[271]
N1	Nε-Acryloyl-L-lysine	0.26	0.06	0.28	[271]
N1	N-Acryloyl-3,6,9-trioxa-1-aminodecane	0.25	0.04	0.35	[271]
N2	A	0.42	0.48	0.69	[272]
N3	A	0.52	0.19	0.53	[272]
N4	A	0.67	0.15	0.71	[272]
N2	N-(2-Hydroxyethyl)-acrylamide	0.26	0.13	0.59	[272]
N3	N-(2-Hydroxyethyl)-acrylamide	0.39	0.27	0.66	[272]
N4	N-(2-Hydroxyethyl)-acrylamide	0.67	0.20	0.63	[272]
N2	N,N-Diethylacrylamide	0.37	0.27	0.51	[272]
N3	N,N-Diethylacrylamide	0.80	0.41	0.56	[272]
N4	N,N-Diethylacrylamide	0.69	0.13	0.61	[272]
N2	N,N-Dimethylacrylamide	0.60	0.55	0.61	[272]
N2	A plus acrylic acid	0.02	0.09	0.40	[272]
N2	A plus 6-methacrylamidohexyl-ammonium chloride	0.33	0.38	0.62	[272]
N2	N-Ethylacrylamide	0.32	0.14	0.67	[272]
N3	N-Ethylacrylamide	0.32	0.24	0.65	[272]
N4	N-Ethylacrylamide	0.61	0.16	0.73	[272]

[a] Abbreviations:

N1, N^6-[N-(N-Acryloyl-1-methoxycarbonyl-5-aminopentyl)-propioamide]-NAD;

N2, N^6-[N-(6-Methacrylamidohexyl)carbamoylmethyl]-NAD;

N3, N^6-[N-[2-[N-(2-Methacrylamidoethyl)carbamoyl]ethyl]carbamoylmethyl-NAD;

N4, N^6-[N-[N-(2-Hydroxy-3-methacrylamidopropyl)carbamoylmethyl]carbamoylmethyl]-NAD.

YADH. These two works triggered a series of investigations on macromolecule-bound NAD, as listed in Tables 2.13 and 2.14.

Two approaches are used for binding nicotineamide coenzymes to macromolecules. In the first approach, NAD is bound to existing macromolecules directly or indirectly via spacers. Investigations with this approach are listed in Table 2.13. In this case, there is a wide choice of macromolecules which can be combined with NAD. In the second approach, a polymerizable NAD-bearing monomer is prepared first and then copolymerized with other monomers. An advantage of this method is the possibility to form a NAD-bearing polymer gel which entraps enzymes. This forms a self-contained system immobilizing all biocatalysts needed, including enzymes and coenzyme. Investigations with this approach are listed in Table 2.14.

ZAPPELI et al. [261] employed the former approach using 3,4-epoxybutanoic acid for the introduction of a side chain into the adenine ring of NAD. The Dimroth rearrangement gave nicotineamide-6-(2-hydroxy-3-carboxypropylamino)-purine dinucleotide, which was bound to either polyethyleneimine or polylysine by using carbodiimide. The coenzyme activity of the final compound was good with LDH (25–60%) but poor with ADH (6–7%) and aldehyde dehydrogenase (2%).

YAMAZAKI et al. [29] employed the direct succinyl-linkage approach at 6-NH$_2$ by succinic anhydride. As the succinyl linkage of NAD had been reported to be labile [260], the stability of that linkage was checked carefully. They concluded that at least 85% of the initial coenzyme activity was retained after dialysis for one week. The relative coenzyme activity of polylysine-NAD compared with native NAD was at around 20% with YADH, LDH and aldehyde dehydrogenase. Polylysine-NAD thus formed was placed in a membrane reactor together with ADH and LDH to form a self-contained system, which could continuously produce lactate with no need for external supply of NAD. The half-life of this system was ten days.

ZAPPELI et al. [262] modified NAD at the C-8 position of adenine. 8-Bromoadenine NAD was prepared first, then reacted with 3-mercaptopropionic acid to form nicotineamide-8-(2-carboxyethylthio)adenine dinucleotide, which could be coupled to polyethylene imine or polylysine. Polyethylene imine-NAD showed a good relative coenzyme activity with ADH (47%) but poor with LDH (3%). The coenzyme activity of polylysine-NAD was poor both with ADH (5%) and with LDH (6%).

SCHMIDT and GRENNER [263] used aziridine to prepare N^6-aminoethyl-NAD, which was coupled to Dextran, Sepharose, and Affi Gel 10. Among these, the relative coenzyme activity of dextran-NAD was from 8 to 37% with YADH, LDH, MDH, and glutamate dehydrogenase, but the coenzyme activities of other NAD derivatives were poor.

FURUKAWA et al. [265] used 3-propiolactone for the first alkylation at N^1 position. After DIMROTH rearrangement, N^6-(2-carboxyethyl)NAD$^+$ was obtained and this was bound to polyethylene glycol with a molecular-weight range from 3000 to 3700. This compound showed good relative coenzyme activities of more than 50% with YADH, HLADH, LDH, and MDH. It became clear that the NAD derivative with the smaller size and with the lower NAD density expressed the higher coenzyme activity.

SAKAGUCHI et al. [266] coupled N^6-aminoethyl-NAD to Dextran T70 via polyglycine as a spacer. By changing the size of polyglycine, they investigated the effect of the length of the spacer. The best results were obtained with spacers having 12–15 atoms in the main chain.

FULLER et al. [267] prepared two types of epoxy-containing polymers for the purpose of chemical modification of NAD: copolymers of 3-[4-(2,3-epoxypropoxy)butoxy]-2-hydroxypropyl acrylate and methacryl choline or N-methacrylyl-2-glucosamine. In the latter case, the epoxide in the polymer was directly reactive with 6-NH$_2$ of NADH. The relative coenzyme activity thus obtained was from moderate to low.

BUCKMANN et al. [268] coupled N^1-(2-aminoethyl)-NAD$^+$ (AE-NAD) to carboxylated polyethylene glycol prior to Dimroth rearrangement. Because of this change in the procedure and the reuse of the unreacted AE-NAD in the coupling reaction process of AE-NAD with polyethylene glycol, the overall NAD-based yield for the final product, polyethylene glycol-N^6-(2-aminoethyl)-NAD, increased up to 68%. The relative coenzyme activity was good with leucine dehydrogenase (87%) and with FDH (210%). In the latter case, the coenzyme activity of the NAD derivative was higher than that of the native one. This NAD derivative gave an excellent NAD cycling number as high as 50 000 when this was used with leucine dehydrogenase from *Bacillus stearothermophilus* and FDH from *Candida boidinii* in a membrane reactor system for the continuous production of leucine. The inactivation rate of this NAD derivative was measured to be 16.8%/day [171].

YOSHIKAWA et al. [269] combined N^6-[(6-aminohexyl)carbamoylmethyl]-NAD$^+$ with casein in a reaction catalyzed by transglutaminase. Thus obtained casein-bound NAD could be recovered for reuse as a precipitate when calcium ion was present.

ADACHI et al. [270] investigated the effect of the size of macromolecules on the coenzyme activity of NAD derivatives. Dextran was selected as a support, the molecular weight of which was changed from 10 000 to 500 000. Unexpectedly, no significant effect of the size of the macromolecule was observed.

On a macromolecule carrying NAD, enzymes may also be coimmobilized. GESTRERIUS et al. [231] coimmobilized both NAD and HLADH on CNBr-activated Sepharose 4B. The immobilized enzyme and coenzyme could interact in the substrate-conjugated oxidoreduction with ethanol and lactaldehyde at an NAD cycling rate of 3400 h^{-1}. The bound NAD, however, could not be regenerated either enzymatically with LDH or chemically with phenazine ethosulfate (coupled to 2,6-dichlorophenol indophenol) or potassium ferricyanide, probably because of the steric hindrance around the bound NAD. MAZID and LAIDLER [239] immobilized NAD and YADH on the surface of a partly hydrolyzed nylon tube. The immobilized NAD could be regenerated chemically with phenazine ethosulfate and 2,6-dichlorophenol indophenol.

As a support macromolecule of the NAD derivative, an enzyme protein itself may be used. N^6-[(6-aminohexyl)-carbamoylmethyl]-NAD was bound to the molecule of HLADH. The bound NAD could be cycled both in a substrate conjugation reaction [233] with ethanol and lactaldehyde at an NAD cycling rate of 40 000 h^{-1} and in a conjugated enzyme reaction with free LDH [234] at an NAD cycling rate of 300 h^{-1}. Immobilization of NAD on the LDH molecule was also attempted [232]. LEGOY et al. coimmobilized NAD with YADH [237] or with steroid dehydrogenase [238] in a crosslinked albumin gel with glutaraldehyde. The immobilized coenzyme was cycled in a conjugation reaction of YADH (reduction of NAD by ethanol) and air oxidation of the coenzyme mediated by phenazine methosulfate. In these investigations on coimmobilization of enzymes and coenzyme, the immobilized NAD on the enzyme protein is expected to act as an artificial prosthetic group of the apoenzyme.

In the second approach for binding the coenzyme to macromolecules, a polymerizable monomer bearing NAD is first prepared, then it is copolymerized with an appropriate counter-monomer. Investigations with this approach are listed in Table 2.14. MURAMATSU et al. [264] prepared NAD$^+$-N^6-[N-(N-acryloyl-1-methoxycarbonyl-5-aminopentyl)-propioamide], which was copolymerized with acrylamide at a good polymerization yield (90%) with a good retention of coenzyme activity. FURUKAWA et al. [271] copolymerized the same monomer bearing NAD with various countermonomers. Among the five countermonomers tested, metha-

132

crylamide gave the best result in terms of coenzyme activity (52–74%) with YADH, LDH, and MDH. An increase in NAD density in the polymer gave the lower coenzyme activity.

YAMAZAKI et al. [272] prepared three different types of NAD-bearing monomers, which were copolymerized with five different countermonomers. Coenzyme activities with various dehydrogenases were from moderate to high as shown in Table 2.14. In these methods, the copolymerization yield of NAD-bearing monomer is fairly good (60–90%). Therefore, the total yield seems to be determined in the first step, i.e., the preparation of NAD-bearing monomer (10–20%).

By using the copolymerization approach, enzymes can also be entrapped in the polymer gel bearing NAD to form a self-contained system for coenzyme cycling. MAEDA et al. [235] prepared N^6-[N-[N-(2-hydroxy-3-methacrylamidopropyl)carbamoylmethyl]carbamoylmethyl]-NAD, which was copolymerized with acrylamide and methylenebisacrylamide to entrap FDH and MDH. The polymer gel thus formed was packed in a column. Under a continuous feed of oxaloacetate and formate, this system produced L-malate for 3 days without external feed of NAD. The overall stability was solely determined by the stability of MDH (pig heart) used. Therefore, this enzyme was replaced by MDH from *Thermus thermophilus* [236]. Then, the activity remained during continuous operation for 3 weeks, with 60% of the initial activity retained even at the final stage. During this period, no loss in activities of immobilized NAD and FDH were observed. However, 95% of the MDH activity was lost again due to leakage. By using this unique approach, a very compact self-contained system may be constructed. A disadvantage of this method exists in the impossibility of reactivating the system when a partial inactivation occurs among the immobilized enzymes and/or coenzyme.

2.4.2 Chemically Modified NADP

Macromolecule-bound NADP derivatives prepared so far are listed in Table 2.15. Macromolecule-bound NADP was first prepared by LOWE et al. [280] using the same technique as for macromolecule-bound NAD [258]. N^6-[N-(6-Aminohexyl)-acetamide]-NADP$^+$ was bound on CNBr-activated Dextran T40. Relative coenzyme activities of Dextran-NADP with G6PDH, 6-phosphogluconate dehydrogenase, and glutamate dehydrogenase were between 10% and 35% but that with isocitrate dehydrogenase was negligible.

ZAPPELI et al. [281] prepared three different types of polyethylene imine-bound NADP. Chemical modifications were made at N-1, N-6, and C-8 positions in the adenine moiety. Unexpectedly, the polyethylene imine-bound NADP derivative at N-1 position gave the highest activity with G6PDH, glutamate dehydrogenase, and aldehyde dehydrogenase. All of the three NADP derivatives, again, did not show any activity with isocitrate dehydrogenase. The inactivity with isocitrate dehydrogenase shows the strict requirement of this enzyme for the structure of the chemically modified coenzyme. OKUDA et al. [283–285] found that the alkylation reaction occurred not only at the N-1 position in adenine but also at 2′ position in ribose when propiolactone was used for the alkylation of NADP. To preclude this, NADP was firstly converted to 2′,3′-cyclic NADP, which was alkylated only

Table 2.15. Derivatives of NADP

Macromolecule	Spacer in adenine ring of NADP	Relative activity with			Ref.
		G6PDH	Gluta-mateDH	Alde-hydeDH	
Dextran T40	N^6-[N-(6-Aminohexyl)-acetamide]	0.35	0.15		[280]
Polyethylene imine	N^1-CH$_2$CH(OH)CH$_2$COOH	0.22	0.23	0.36	[281]
Polyethylene imine	N^6-CH$_2$CH(OH)CH$_2$COOH	0.16	0.03	0.29	[281]
Polyethylene imine	8-S(CH$_2$)$_2$COOH	0.10	0.02	0.01	[281]
Polyethylene glycol	N^6-(CH$_2$)$_2$COOH	0.69		0.16	[283–285]
Copolymer of N^6-[N-[2-[N-(2-methacrylamido-ethyl)carbamoyl]ethyl]carbamoylmethyl]-NADP (I) and acrylamide		1.16	0.30		[286]
Copolymer of I and acrylamide plus N,N-dimethyl-N-(2-methacrylamidoethyl)-ammonium chloride		0.29	0.10		[286]
Copolymer of I and acrylamide plus sodium acrylate		0.26	0.27		[286]

at the N-1 position in adenine, converted to the N^6 derivative through Dimroth rearrangement, and bound to polyethylene glycol (PEG). Thus prepared PEG-bound 2′,3′-cyclic NADP was enzymatically hydrolyzed to PEG-NADP by 2′,3′-cyclic-nucleotide 3′-phosphodiesterase. The coenzyme activity of PEG-NADP was good with G6PDH, 6-phosphogluconate dehydrogenase, and glutamate dehydrogenase (69–95%), moderate with aldehyde dehydrogenase (16%), and poor with isocitrate dehydrogenase (0.02%). OKUDA et al. [285] also found that NADP and its derivatives treated with carbodiimide were converted to 2′,3′-cyclic derivatives. This may explain the higher coenzyme activity of the NADP derivative by them as compared to the results obtained by other groups [258, 281] because 2′,3′-cyclic NADP is almost inactive as a coenzyme.

ARAKI et al. [286] prepared polymer-bound NADP derivatives by the copolymerization approach. A polymerizable NADP derivative, N^6-[N-[2-[N-(2-methacrylamidoethyl)carbamoyl]-ethyl]carbamoylmethyl]-NADP was synthesized first. This NADP derivative was copolymerized with three different countermonomers: (1) acrylamide, (2) acrylamide plus N,N-dimethyl-N-(2-methacrylamidoethyl)ammonium chloride, and (3) acrylamide plus sodium acrylate. The last two polymers were positively and negatively charged, respectively, while the first polymer was uncharged. The coenzyme activity of polymer-bound NADP was best for the uncharged polymer.

2.4.3 Chemically Modified ATP and Others

Investigations on macromolecule-bound ADP/ATP are listed in Tables 2.16 and 2.17. FULLER and BRIGHT [287] prepared an epoxy-containing acrylic copolymer of butanediolglycidylglycerylether acrylate with methacrylyl choline, which could directly react with ADP/ATP at 6-NH$_2$ in the adenine moiety. The ADP/ATP de-

Table 2.16. Derivatives of ADP/ATP

Macromolecule	Spacer in adenine ring	Relative activity with			Ref.
		HK[a]	Glyce-rolK[b]	AcK[c]	
Copolymer of (**I**) and methacryl choline	Butanediolglycidylglyceryl ether acrylate (**I**) bound at N^6	~1		0.20	[287]
Dextran T40	$N^6HCONH(CH_2)_2NH_2$	0.49	0.70	0.29	[288]
Dextran T40	$N^6HCH_2S(CH_2)_2CONH(CH_2)_3$-$-NH_2$	0.58	0.92	0.35	[289]
Dextran T40	$N^6[CH_2S(CH_2)_2CONH$-$-(CH^2)_3NH_2]_2$	0.49	0.58	0.15	[289]

[a] Hexokinase.
[b] Glycerol kinase.
[c] Acetate kinase.

Table 2.17. ATP/ADP-bearing copolymers [290]

ATP/ADP bearing monomer[a]	Monomer to be copolymerized	Relative activity with·	
		HK	GlycerolK
A1	Acrylamide	0.66	0.32
A2	Acrylamide	0.50	0.30
A3	Acrylamide	0.50	0.26
A1	N-(2-Hydroxyethyl)acrylamide	0.46	–
A2	N-(2-Hydroxyethyl)acrylamide	0.42	0.21
A3	N-(2-Hydroxyethyl)acrylamide	0.40	0.25
A1	N-Ethylacrylamide	0.53	0.30
A2	N-Ethylacrylamide	0.55	0.26
A3	N-Ethylacrylamide	0.60	0.26
A1	N,N-Diethylacrylamide	0.35	0.22
A2	N,N-Diethylacrylamide	0.21	0.24
A3	N,N-Diethylacrylamide	0.21	0.21
A2	Acrylamide plus acrylic acid	0.13	0.27
A2	Acrylamide plus 6-methacrylamidohexyl ammonium chloride	0.59	0.31

[a] Abbreviations:
A1, N^6-[N-(6-Methacrylamidohexyl)carbamoylmethyl]-ATP;
A2, N^6-[N-[2-[N-(2-Methacrylamidoethyl]carbamoyl]ethyl]carbamoylmethyl]-ATP;
A3, N^6-[N-[N-(2-Hydroxy-3-methacrylamidopropyl)carbamoylmethyl]carbamoylmethyl]-ATP.

rivative obtained showed good coenzyme activity with hexokinase and acetate kinase. YAMAZAKI et al. [288, 289] prepared three different ADP/ATP derivatives: N^6-[N-(6-aminohexyl)carbamoyl]-, N^6-[[N-(3-aminopropyl)carbamoylethyl]thio-methyl]-, and N^6,N^6-bis[[N-(3-aminopropyl)carbamoylethyl]thiomethyl-ADP/ATP. Hexamethylene diisocyanate was used for the preparation of the first derivative and formaldehyde with 3-mercaptopropionic acid was used for the second and the third derivative. These compounds were bound to CNBr-activated Dextran T40. Activities of these macromolecule-bound ADP/ATP derivatives were good with hexokinase and glycerol kinase, moderate with acetate kinase, and poor with phosphoglycerate kinase and pyruvate kinase.

YAMAZAKI and MAEDA [290] employed the copolymerization approach to prepare polymer-bound ATP. Three different polymerizable ATP derivatives were first prepared: N^6-[N-(6-methacrylamidohexyl)carbamoylmethyl]-, N^6-[N-[2-[N-(2-methacrylamidoethyl)carbamoyl]ethyl]carbamoylmethyl]-, and N^6-[N-[N-(2-hydroxy-3-methacrylamidopropyl)carbamoylmethyl]carbamoylmethyl]-ATP. These ATP derivatives were copolymerized with six different countermonomers: acrylamide, N-(2-hydroxyethyl)acrylamide, N-ethylacrylamide, N,N-diethylacrylamide, acrylamide plus acrylic acid, and acrylamide plus 6-methacrylamidohexyl ammonium chloride. Most of the polymer derivatives of ATP thus prepared showed good coenzyme activities with hexokinase (35–66%) and with glycerol kinase (25–32%).

The techniques described here for preparing NAD-, NADP-, and ATP derivatives were also applied to prepare derivatives of other coenzymes such as FAD [291] and CoA [292, 293].

2.5 Bioreactor System with Continuous Coenzyme Cycling

2.5.1 Reactor System Design

For a large-scale production, continuous operation should be considered. To this end, the first problem to be solved is how the main reaction (coenzyme-consuming reaction) is combined with the subreaction (coenzyme-regeneration reaction) involving the separator for coenzyme recovery. There are two possible choices: (1) consumption-recovery-regeneration approach (Fig. 2.14 A) and (2) in situ regeneration approach (Fig. 2.14 B). In the consumption-recovery-regeneration approach, the coenzyme needed is consumed first in a main reactor, recovered at a separator, and regenerated to the original active state in a subreactor. In this approach, the total cycling number of the coenzyme is completely determined by the separation efficiency at the separator. Even if 99% recovery (not easily realized) for coenzyme is assumed, the cycling number of the coenzyme still is at most 100. On the other hand, in the in situ regeneration approach, no restriction exists in the cycling number of coenzyme, in principle, provided that a sufficient amount of substrate is available and that coenzyme, free or immobilized, is sufficiently stable. In addition, an unfavorable equilibrium of the main reaction may be shifted toward a favorable direction through combination with the regeneration reaction. Also, a separator for the recovery of coenzyme may be installed at the outlet for improving the coenzyme cycling number. Therefore, the superiority of the in situ approach is evident whenever the two conjugated reactions are compatible in their optimum pH and temperature.

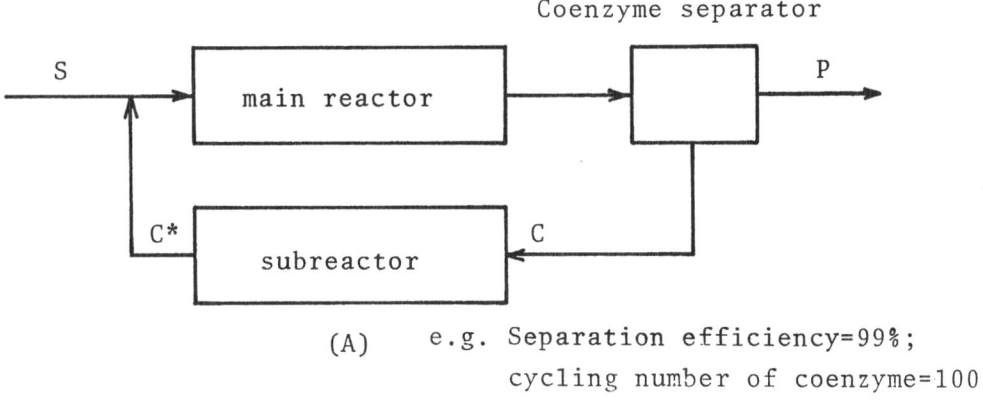

Coenzyme separator

S

main reactor

P

C*

subreactor

C

(A) e.g. Separation efficiency=99%;
cycling number of coenzyme=100

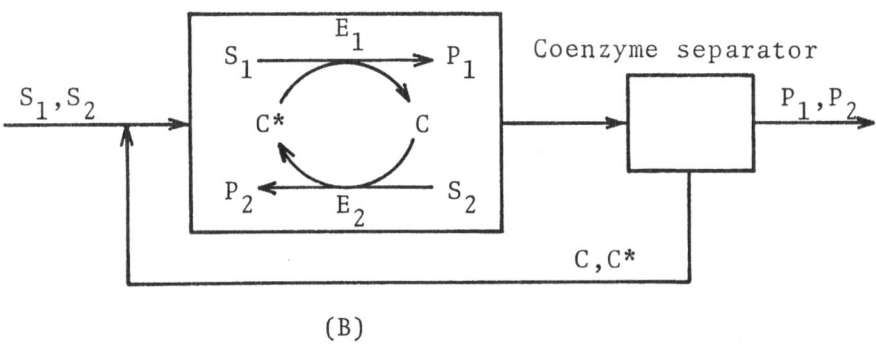

S_1, S_2

$S_1 \xrightarrow{E_1} P_1$

C*   C

$P_2 \xleftarrow{E_2} S_2$

Coenzyme separator

P_1, P_2

C, C^*

(B)

Fig. 2.14. Bioreactor system with coenzyme cycling

Then the next problem to be solved is what type of immobilized enzymes and coenzymes are employed to constitute a bioreactor system. Possible combinations are classified into 12 categories depending on the state of the enzymes and coenzyme as listed in Table 2.18. In categories (1)–(9) in this table, the coenzyme is immobilized together with the enzymes for cycling reaction so that there is no need for external feed of the coenzyme in a continuous operation. This is considered as a "self-contained" system, which operates continuously under the feed of substrates for the main reaction and for the coenzyme regeneration. Depending on the state of the enzymes and the coenzyme, the secondary entrapment of these biocatalysts in a gel, in a microcapsule, or in a membrane reactor system should be considered for completion of the self-containment. On the other hand, enzymes only are immobilized in categories (10)–(12) and free coenzyme is externally supplied together with substrates. These are considered as "partially-contained" systems.

Category (1) forms a complete self-contained system with no need for secondary entrapment. Enzymes and coenzyme should be immobilized on the same solid support to ensure mutual interaction. GESTRERIUS et al. [231] attempted this approach with ADH and NAD both immobilized on CNBr-activated Sepharose 4B.

137

Table 2.18. Classification of continuous coenzyme cycling system

Category	Need for feed of coenzyme	Coenzyme bound on	Enzyme bound on	Need for secondary entrapment
1	N	Solid support	Solid support	N
2	N	Solid support	Macromolecule	Y
3	N	Solid support	Native	Y/N
4	N	Macromolecule	Solid support	Y
5	N	Macromolecule	Macromolecule	Y
6	N	Macromolecule	Native	Y
7	N	Native	Solid support	Y
8	N	Native	Macromolecule	Y
9	N	Native	Native	Y
10	Y	Native	Solid support	N
11	Y	Native	Macromolecule	Y
12	Y	Native	Native	Y

The immobilized NAD functioned with the immobilized ADH but could not interact with externally added free LDH for the regeneration reaction probably because of the steric hindrance around the immobilized NAD. The same problem would also be expected in categories (2) and (4), although no investigations have been reported in these categories so far.

As for category (3), MAEDA et al. [235, 236] prepared a polymerizable monomer-bearing NAD. Enzymes needed were entrapped in a copolymer gel containing the monomer-bearing NAD. This formed a self-contained system for coenzyme cycling with no need for secondary entrapment. Details of this have been already given in Sect. 2.4.1.

Category (6) includes methods with water-soluble macromolecule-bound coenzymes and have been investigated most extensively. In this case, the macromolecule-bound coenzyme, together with enzymes for the cycling reaction, is entrapped in a gel, in a microcapsule, or in a membrane reactor system. Among these, the membrane reactor system is most conveniently used because of the ease of immobilization and the possibility for reactivation of the system simply by adding fresh coenzyme and/or enzymes when needed. This system gives a substantially high cycling number of coenzyme and a good productivity of the reactor when appropriately operated [173–176].

In categories (7)–(9), the native coenzyme is physically entrapped together with enzymes, native or immobilized. For this purpose, a very strictly selective membrane should be chosen because the molecular weights of native coenzymes are between 500 and 800. There are no investigations so far in categories (7) and (8). As for category (9), CHAMBERS et al. [64] used a reverse osmosis-type membrane with a cut-off molecular weight of 200. MAY and LANDGRAFF [51] used hydrocarbon-based liquid membrane microcapsules for the immobilization of native NAD together with enzymes. In these cases, the poor permeability of substrates and products through the membrane is another problem to be solved because of the tightness of the membrane used.

In categories (10)–(12), coenzymes are not immobilized but supplied externally to the reactor containing immobilized enzymes. This is not a self-contained system but is a "partially-contained" system. This is a technically easier approach as compared with the more sophisticated self-contained case especially as in categories (1) and (6) because no chemical modification of coenzyme is involved. In this case, there is no problem with the activity and the specificity of coenzyme because native coenzyme itself is used. The stability problem of coenzyme could be also circumvented because coenzyme is fed continuously. A problem in this case exists in the cycling number of coenzyme because there is no immobilization. The cycling number of coenzyme is determined from the reactor kinetics discussed in Sect. 2.5.2. Depending on the concentration of immobilized enzymes, an affinity interaction between the enzymes and coenzyme would be expected. This occurs when the enzyme concentration is very high. When this effect exists, the mean residence time of coenzymes in a reactor will be much longer than for other substrates (and products) as if the coenzyme is "dynamically immobilized" on the enzymes in the reactor. Under this condition, the cycling number of the coenzyme goes up very high. More details on this will be given later in Sect. 2.5.3.

2.5.2 Reactor Kinetics

A theoretical model for a continuously operated reactor in category (6) in Table 2.18, which may be considered as a representative case of the self-contained systems in categories (1)–(9), was studied by WICHMANN and WANDREY [171] and also by KATAYAMA et al. [319]. The results were compared with the experimental results.

WICHMANN and WANDREY [171] studied a membrane reactor system with L-leucine dehydrogenase/FDH/polyethylene glycol-bound NAD for the production of L-leucine from α-ketoisocaproate. To describe the steady-state behavior of the reactor, they considered only the forward reactions, neglecting the backward reactions in their theoretical model. This assumption seemed appropriate in their case. The theoretical predictions well agreed with the experimental results.

KATAYAMA et al. [319] constructed a more complete theoretical model for the steady-state behavior of a membrane reactor with LDH/ADH/polyethylene-bound NAD considering backward reactions. At the steady state, the material balances are expressed as follows:

$$v_{LDH} = -v_{ADH},\tag{2.14}$$

$$v_{LDH} = [\text{lactate}]/\tau,\tag{2.15}$$

$$[\text{pyruvate}]_0 = [\text{pyruvate}] + [\text{lactate}],\tag{2.16}$$

$$[\text{ethanol}]_0 = [\text{ethanol}] + [\text{acetaldehyde}],\tag{2.17}$$

$$[\text{lactate}] = [\text{acetaldehyde}],\tag{2.18}$$

$$[\text{PEG-NAD}]_0 = [\text{PEG-NAD}^+] + [\text{PEG-NADH}],\tag{2.19}$$

where τ is the mean residence time and the subscript "0" corresponds to the initial state or the state before entering into the reactor. The enzyme reaction rates, v_{LDH}

and v_{ADH}, based on a simplified ordered bi-bi mechanism were expressed as follows:

$$v_{LDH} = num_1/den_1 \,, \tag{2.20}$$

$$v_{ADN} = num_2/den_2 \,, \tag{2.21}$$

$$num_1 = V_f V_r [LDH]([PEG-NAD]\,[lactate]$$
$$-[PEG-NADH]\,[pyruvate]/K_{eq}) \tag{2.22}$$

$$den_1 = V_r K_{iNAD} K_{lactate} + V_r K_{lactate}[PEG-NAD]$$
$$+ V_r K_{NAD}[lactate] + V_r[PEG-NAD]\,[lactate]$$
$$+ V_f(K_{NADH}[pyruvate] + K_{pyruvate}[PEG-NADH]$$
$$+ [PEG-NADH]\,[pyruvate])/K_{eq}$$
$$+ V_f K_{NADH}[PEG-NAD]\,[pyruvate]/K_{eq}K_{iNAD}$$
$$+ V_r K_{NAD}[PEG-NADH]\,[lactate]/K_{iNADH} \,, \tag{2.23}$$

$$num_2 = V_f V_r [ADH]([PEG-NAD]\,[ethanol]$$
$$- [PEG-NADH]\,[acetaldehyde]/K_{eq}) \,, \tag{2.24}$$

$$den_2 = V_r K_{iNAD} K_{ethanol} + V_r K_{ethanol}[PEG-NAD] + V_r K_{NAD}[ethanol]$$
$$V_r[PEG-NAD]\,[ethanol] + V_f(K_{NADH}[acetaldehyde]$$
$$+ K_{acetaldehyde}[PEG-NADH]$$
$$+ [PEG-NADH]\,[acetaldehyde])/K_{eq}$$
$$+ V_f K_{NADH}[PEG-NAD]\,[acetaldehyde]/K_{eq}K_{iNAD}$$
$$+ V_f K_{NAD}[PEG-NADH]\,[ethanol]/K_{iNADH} \,. \tag{2.25}$$

This system with six unknowns was numerically solved and the effect of the operating conditions on the overall behavior of the reactor were calculated with various concentrations of enzymes, coenzyme, and substrates. A general agreement was obtained between the theoretical calculations and the experimental results.

The theoretical model described above is also applicable to the case with a "partially-contained" system with continuous feed of coenzyme as the categories (9)–(12) in Table 2.18 if Eq. (2.19) is replaced with the following equation:

$$[NAD]_0 = [NAD^+] + [NADH], \tag{2.26}$$

where $[NAD]_0$ is the concentration of the coenzyme supplied continuously from the inlet of the reactor. The enzyme kinetic models in Eqs. (2.20) to (2.25), however, are applicable only to systems with neglegible concentrations of enzyme-coenzyme complexes compared with the total coenzyme existing in the reactor.

140

When the concentrations of enzyme-coenzyme complexes are not neglegible, more precise mathematical model would be needed. The authors [146, 321] investigated a system with LDH/ADH being physically immobilized in an ultrafiltration hollow fiber tube, which was placed in a complete-mixing type reactor as shown in Fig. 2.15. The free coenzyme, NAD, was continuously fed from the inlet. In this case, the concentrations of immobilized enzymes in the hollow fiber was substantially higher compared with the coenzyme fed continuously. The theoretical model for this is schematically expressed in Fig. 2.16 with the assumption of Theorell-Chance mechanism for the enzyme reactions. The reaction from pyruvate to lactate catalyzed by LDH was considered to be the main reaction. The material balances for every possible intermediates are:

$$[O]_f + [R]_f = [O]^* + [R]^* , \tag{2.27}$$

$$F([O]_f - [O]) = K_N S([O] - [O]^*) , \tag{2.28}$$

$$F([R]_f - [R]) = K_N S([R] - [R]^*) , \tag{2.29}$$

$$[Pyr]_f + [Lac]_f = [Pyr]^* + [Lac]^* , \tag{2.30}$$

$$F([Lac]_f - [Lac]) = K_S S([Lac] - [Lac]^*) , \tag{2.31}$$

$$K_N S([O] - [O]^*) = V_F(k_1[A]^*[O]^* + k_7[L]^*[O]^* - k_2[AO]^* - k_8[LO]^*) , \tag{2.32}$$

$$K_N S([R] - [R]^*) = V_F(k_6[A]^*[R]^* + k_{12}[L]^*[R]^* - k_5[AR]^* - k_{11}[LR]^*) , \tag{2.33}$$

$$k_2[AO]^* + k_5[AR]^* = k_1[A]^*[O]^* + k_6[A]^*[R]^* , \tag{2.34}$$

$$[A]^* + [AO]^* + [AR]^* = A^i , \tag{2.35}$$

$$[L]^* + [LO]^* + [LR]^* = L^i , \tag{2.36}$$

$$k_7[L]^*[O]^* + k_{10}[LR]^*[Pyr]^* - k_9[LO]^*[Lac]^* - k_8[LO]^* = 0 , \tag{2.37}$$

$$K_S S([Pyr] - [Pyr]^*) = V_F(k_{10}[LR]^*[Pyr]^* - k_9[LO]^*[Lac]^*) , \tag{2.38}$$

$$F([Pyr]_f - [Pyr]) = K_S S([Pyr] - [Pyr]^*) , \tag{2.39}$$

$$F([Et]_f - [Et]) = K_S S([Et]_f - [Et]^*) , \tag{2.40}$$

$$[Et]_f + [Ald]_f = [Et] + [Ald] , \tag{2.41}$$

$$K_S S([Et] - [Et]^*) = V_F(k_3([AO]^*[Et]^* - k_4[AR]^*[Ald]^*) , \tag{2.42}$$

$$k_1[A]^*[O]^* + k_4[AR]^*[Ald]^* - k_3[AO]^*[Et]^* - k_2[AO]^* = 0 , \tag{2.43}$$

$$[Et]_f + [Ald]_f = [Et]^* + [Ald]^* , \tag{2.44}$$

where O, R, A, and L stand for NAD^+, NADH, ADH, and LDH, respectively. F is the flow rate, S is the surface area of the hollow fiber, and V_F is the volume

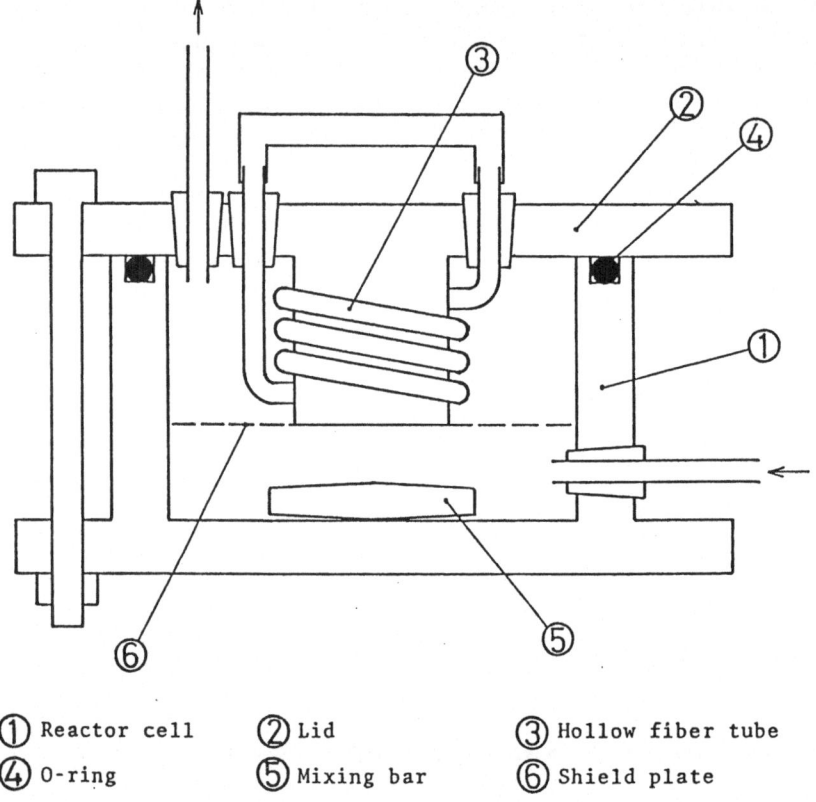

① Reactor cell   ② Lid   ③ Hollow fiber tube

④ O-ring   ⑤ Mixing bar   ⑥ Shield plate

Fig. 2.15. A complete-mixing type reactor containing hollow fiber in which enzymes are entrapped

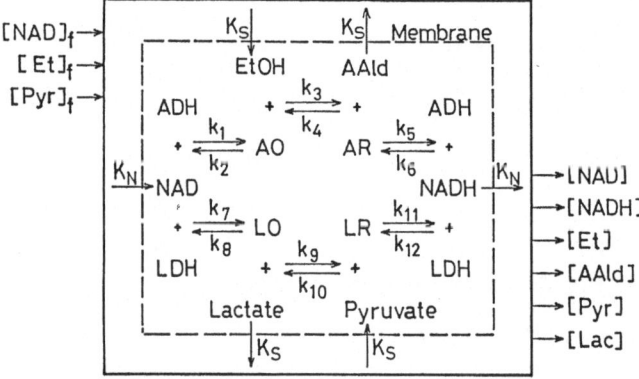

Fig. 2.16. Scheme of continuous NAD cycling by conjugated enzyme system of alcohol dehydrogenase (ADH) and lactate dehydrogenase (LDH) immobilized in ultrafiltration hollow fiber

Table 2.19. Reactor conditions and kinetic parameters

Reactor conditions:	$V = 30$ cm^2, $\alpha = 0.03$, $r = 0.04$ cm $K_N = 5.74 \times 10^{-5}$ cm/s, $K_S = 2.27 \times 10^{-4}$ cm/s
Kinetic parameters:	(pH $= 8$, 25 °C)
Yeast ADH:	$k_1 = 4.65 \times 10^6$, $k_2 = 2670$, $k_3 = 4.45 \times 10^4$ $k_4 = 2.00 \times 10^6$, $k_5 = 555.6$, $k_6 = 2.04 \times 10^7$
Bovine heart LDH:	$k_7 = 1.34 \times 10^6$, $k_8 = 216.7$, $k_9 = 1.52 \times 10^4$ $k_{10} = 1.47 \times 10^6$, $k_{11} = 88.3$, $k_{12} = 1.08 \times 10^7$

Table 2.20. Typical example

Operating conditions:	$F = 29.1$ cm^3/h, $A^i = L^i = 1.0 \times 10^{-4}$ mol/l $[O]_f = 1.0 \times 10^{-6}$ mol/l, $[Et]_f = 1.00$, $[Pyr]_f = 0.05$
Results:	$[A]^* = 9.957 \times 10^{-5}$, $[AO]^* = 1.44 \times 10^{-7}$, $[AR]^* = 2.87 \times 10^{-7}$ $[L]^* = 9.916 \times 10^{-5}$, $[LO]^* = 8.35 \times 10^{-7}$, $[LR]^* = 3.38 \times 10^{-9}$ $[O]^* = 9.486 \times 10^{-7}$, $[R]^* = 5.14 \times 10^{-8}$, $[Et]^* = 0.989$ $[Ald]^* = 1.10 \times 10^{-2}$, $[Pyr]^* = 3.90 \times 10^{-2}$, $[Lac]^* = 1.10 \times 10^{-2}$ $[O] = 9.876 \times 10^{-7}$, $[R] = 1.24 \times 10^{-8}$, $[Et] = 0.994$ $[Ald] = 6.11 \times 10^{-3}$, $[Pyr] = 4.39 \times 10^{-2}$, $[Lac] = 6.11 \times 10^{-3}$ $R_N = 6114$ cycle

of the hollow fiber. Concentrations with the suffix "f" and the asterisk "*" corresponds to those in the feed stream and those at the inside of the hollow fiber, respectively. Concentrations without suffix or superscript correspond to those at the outside of the hollow fiber in the reactor. K_N and K_S are the permeabilities through the membrane for the coenzyme and for the other components including substrates and products, respectively.

These equations give a simultaneous system with 18 unknowns, which could be solved numerically by the trial-and-error method. The reactor conditions and the kinetic parameters are listed in Table 2.19, where V is the reactor volume, r is the radius of the hollow fiber, and α is the volume ratio of the hollow fiber in the reactor. A typical theoretical result is shown in Table 2.20. At the operating conditions in Table 2.20, the concentrations of immobilized enzymes are much higher than that of coenzyme fed continuously. It is well understood that the concentrations of enzyme-coenzyme complexes such as $[AO]^*$, $[AR]^*$, $[LO]^*$, and $[LR]^*$ cannot be neglected compared with those of free coenzyme ($[O]^*$ and $[R]^*$) in the immobilized enzyme phase (inside of the hollow fiber). The total concentration of NAD in the immobilized enzyme phase, T_{NAD}, is:

$$T_{NAD} = [O]^* + [R]^* + [AO]^* + [AR]^* + [LO]^* + [LR]^*$$
$$= 2.27 \times 10^{-6} \, M. \tag{2.45}$$

This value is substantially higher than the concentration of NAD fed from the inlet (1.0×10^{-6} M). This shows that NAD is locally concentrated in the immobi-

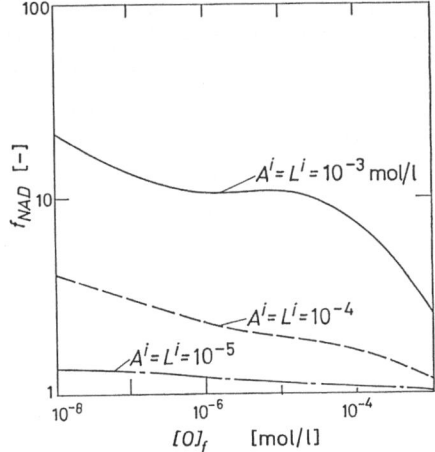

Fig. 2.17. Concentration factor of NAD under the conditions in Tables 2.19 and 2.20

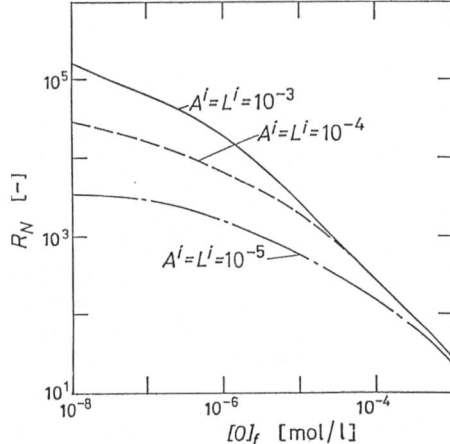

Fig. 2.18. Effect of feed NAD concentration and immobilized enzyme concentration on cycling number of NAD under the conditions in Tables 2.19 and 2.20

lized enzyme phase through the affinity interaction between enzyme and coenzyme. A concentration factor of coenzyme through the affinity, f_{NAD}, is defined as:

$$f_{NAD} = T_{NAD}/([O]_f + [R]_f). \tag{2.46}$$

In Fig. 2.17, f_{NAD} is plotted against $[O]_f$ ($[R]_f$ assumed to be zero) at three different enzyme concentration levels. Under the conditions assumed here, f_{NAD}, being much higher than unity, increased with an increase in the enzyme concentration and/or a decrease in the coenzyme concentration. The greater value of f_{NAD} than unity shows the expression of the "dynamic immobilization". Under the same conditions, the cycling number of NAD, R_N, was calculated and shown in Fig. 2.18. At the higher enzyme concentration and/or at the lower coenzyme concentration, R_N also became higher. Figure 2.18 shows that R_N of 1000 or more

144

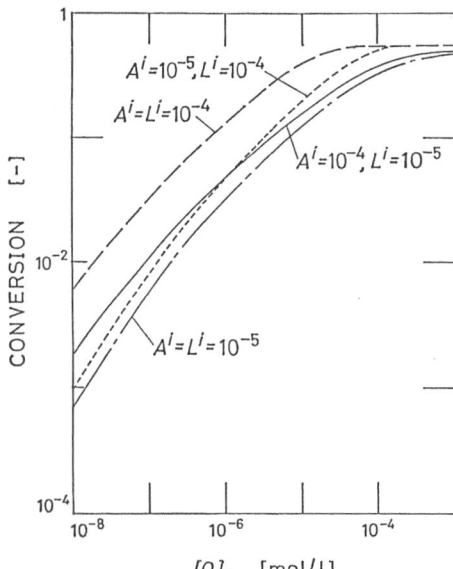

Fig. 2.19. Effect of feed NAD concentration and immobilized enzyme concentration on conversion of pyruvate under the conditions in Tables 2.19 and 2.20

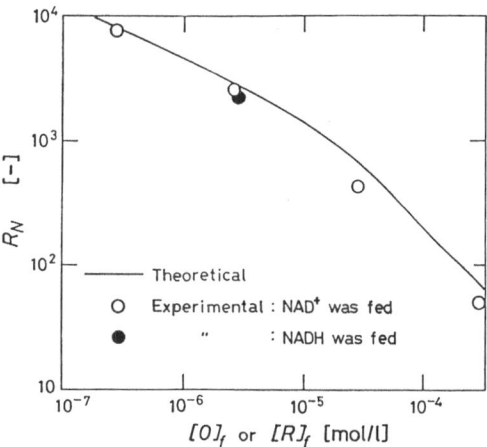

Fig. 2.20. Effect of feed NAD concentration on NAD cycling number: comparison of theoretical model with experimental results. ($A^i = 7.98 \times 10^{-5}$ mol/l, $L^i = 1.33 \times 10^{-4}$ mol/l, $\tau = 0.993$ h)

with a conversion (shown in Fig. 2.19) over 50% is easily possible in the present case. These theoretical results well agreed with the experimental results as shown in Fig. 2.20 [146]. These facts show that the "partially-contained" approach with no immobilization of coenzyme may be practical in some situations because of its simplicity or easiness in practise.

2.5.3 Affinity Chromatographic Reactor

The concept of the "dynamic immobilization" through the affinity has become clear from the discussions given above. The authors tried to extend this idea fur-

145

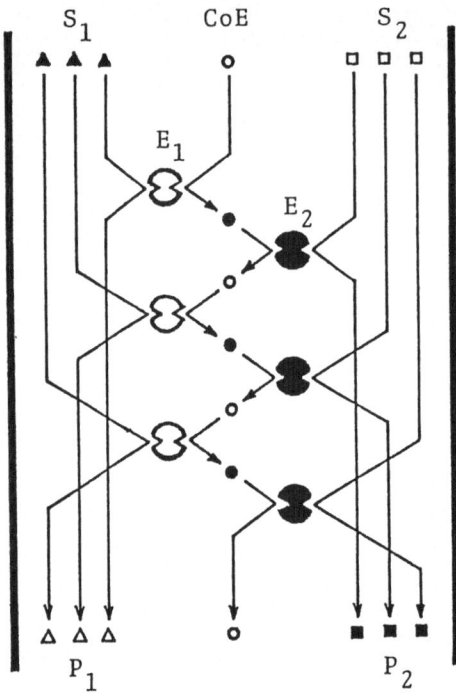

Fig. 2.21. "Dynamic affinity" between a free coenzyme and immobilized conjugated enzymes in affinity chromatographic reactor. (Abbreviations: CoE, coenzyme; E_1 and E_2, enzyme; S_1 and S_2, substrates; P_1 and P_2, products)

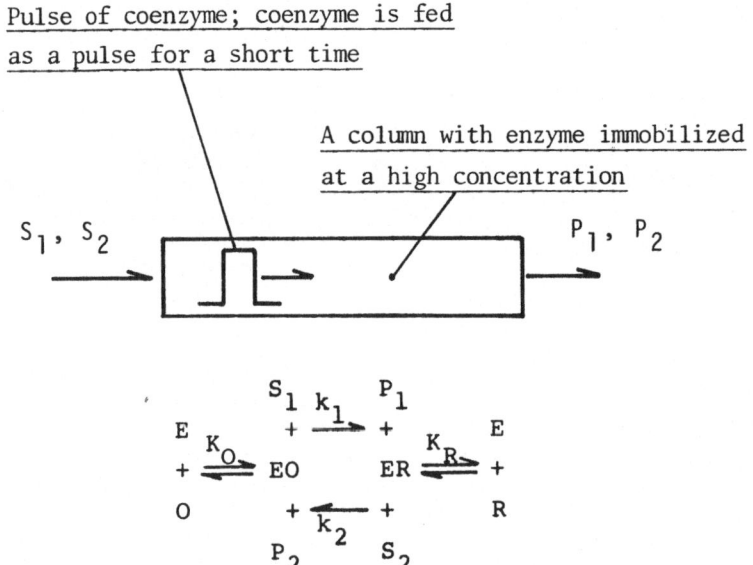

Fig. 2.22. Reaction scheme of coenzyme cycling in an affinity chromatographic reactor with a single enzyme immobilized

ther. For this purpose, the complete-mixing-type reactor is replaced with a col-umn-type reactor. Conjugated enzymes for coenzyme cycling are immobilized at a very high concentration level in the reactor. To this, the conjugated substrates for the main reaction and for the coenzyme regeneration are fed continuously. Then, a required coenzyme is supplied as a pulse for a short time. Because of the high enzyme concentrations in the column, the coenzyme molecule repeats bind-ing and dissociation with immobilized enzymes many times in the cycling reaction before it leaves the reactor, as shown in Fig. 2.21. Thus, a very long residence time and a very high turnover or cycling number of the coenzyme would be expected through the "dynamic affinity". A reactor of this type may be called as an affinity chromatographic reactor.

The simplest-type affinity chromatographic reactor would be composed of a single bifunctional enzyme sujch as ADH [141], which catalyzes both reactions in the coenzyme cycling. Here, the cycling reaction occurs only at the part of the reactor where the coenzyme pulse exists, as shown in Fig. 2.22. In this figure, E, O, and R stand for the immobilized single bifunctional enzyme (ADH) and the two states (oxidized and reduced) of the coenzyme (NAD), respectively. K_O and K_R are the dissociation constants of the enzyme-coenzyme complexes for the dy-namic state. S_1 and S_2 are the substrates for the main reaction and the regener-ation reaction. In the theoretical model, Theorell-Chance mechanism was as-sumed and the backward reaction was neglected. Under steady-state conditions, the system equations are expressed as follows:

$$k_1[EO] = k_2[ER], \tag{2.47}$$

$$K_O = [E][O]/[EO], \tag{2.48}$$

$$K_R = [E][R]/[ER], \tag{2.49}$$

$$E_i = [E] + [EO] + [ER], \tag{2.50}$$

$$O_i + R_i = [O] + [R] + [EO] + [ER], \tag{2.51}$$

where the subscript "i" corresponds to the initial state or the state before the feed.

These nonlinear equations are solved numerically and the mean residence time (τ_C) of the coenzyme fed as a pulse is given as follows:

$$\tau_C/\tau_0 = 1 + D_C, \tag{2.52}$$

$$D_C = ([EO] + [ER])/([O] + [R]), \tag{2.53}$$

where τ_0 is the residence time of compounds having no affinity interaction with the immobilized enzymes.

For the simplest case with equal dissociation constants for O and R, theoretical results are shown in Fig. 2.23, where the greater τ_C/τ_0 means the greater affinity interaction between the enzyme and the coenzyme. From this figure, the necessary condition for sufficient expression of the affinity interaction will be:

$$(O_i + R_i) < (K_O \text{ and } K_R) < E_i. \tag{2.54}$$

147

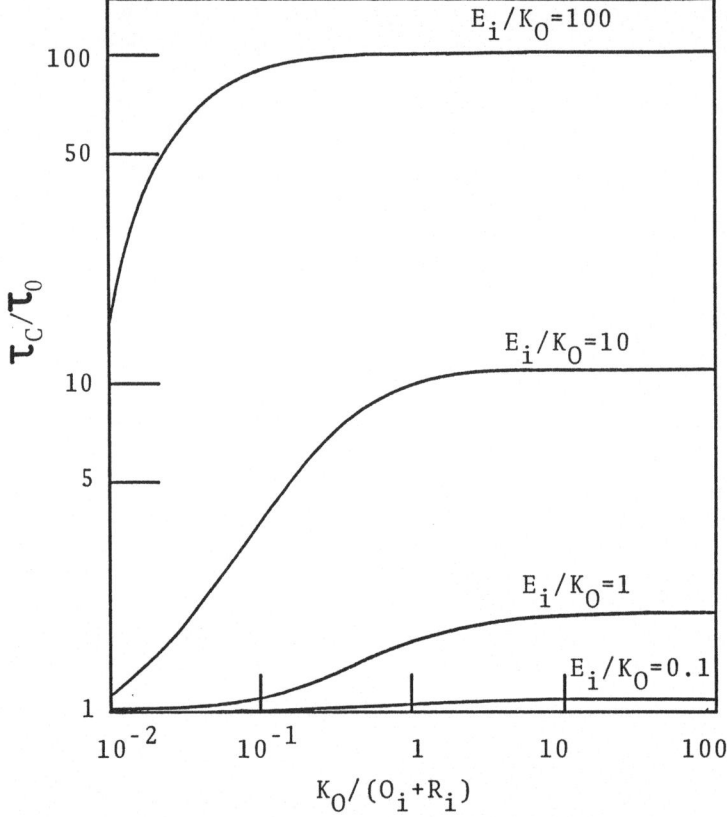

Fig. 2.23. Retention time of coenzyme pulse in an affinity chromatographic reactor with a single enzyme immobilized; $K_R/K_O = 1$; k_2/k_1 was varied from 0.01 to 100

With this condition, τ_C/τ_0 will be larger than unity because of the "dynamic immobilization" of the coenzyme on the immobilized enzymes.

To test the theoretical results described above, we used a ultrafiltration-hollow-fiber capillary reactor as shown in Fig. 2.24. The enzyme (YADH) was physically immobilized in an ultrafiltration hollow-fiber tube, which was inserted in a fine silicon or a glass tube. This reactor has a large volume ratio for the immobilized enzyme phase and can be considered as a column-type reactor provided that the residence time is long enough to allow the coenzyme, the substrates, and the products to permeate freely through the ultrafiltration hollow-fiber membrane, the cut-off molecular weight of which is 13 000. To this reactor, a buffered solution of propionaldehyde and ethanol was fed continuously. After equilibration, NAD was supplied as a pulse for 40 min. The observed value of τ_C/τ_0 is listed in Table 2.21. The dissociation constants for yeast ADH estimated from Michaelis constants are [28]:

$$K_O = 1.18 \times 10^{-4}, \quad K_R = 1.30 \times 10^{-4} \; M.$$

148

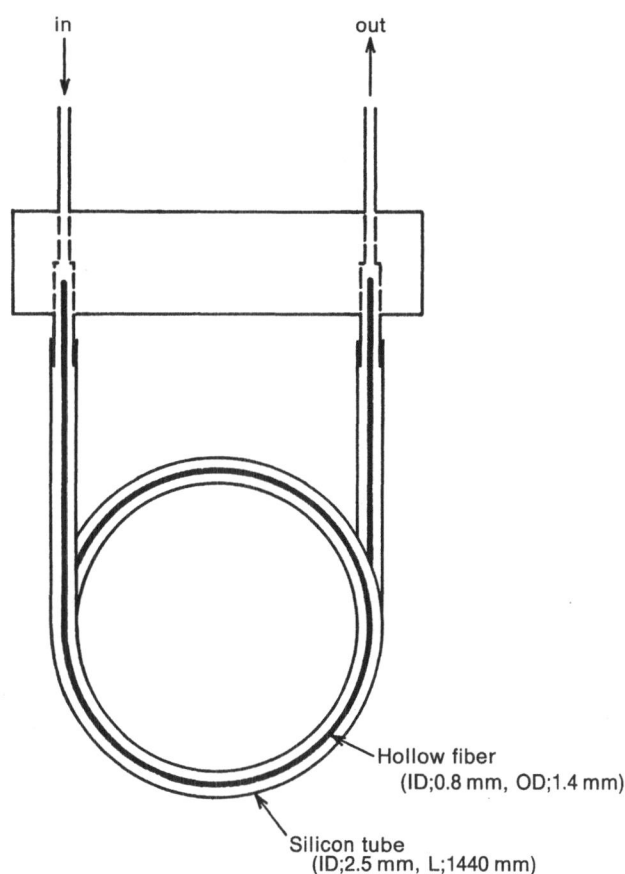

Fig. 2.24. A hollow-fiber-capillary reactor

Hollow fiber
(ID;0.8 mm, OD;1.4 mm)

Silicon tube
(ID;2.5 mm, L;1440 mm)

Table 2.21. Observed nondimensional retention time (τ_C/τ_0) of NAD in affinity chromatographic reactor

C_{ADH} \\ C_{NAD}	1.0×10^{-4}	5.0×10^{-4}	2.5×10^{-3}	1.25×10^{-2}
1.53×10^{-5}	1.53	1.24	–	–
4.82×10^{-5}	1.70	1.36	1.10	–
1.40×10^{-4}	3.20	2.75	1.59	–
3.85×10^{-4}	VL[a]	5.06	2.66	1.39

[a] Very large.

These values and Eq. (2.54) predict that a strong affinity effect will be expected when the concentration level of the immobilized enzyme is higher than 10^{-4} M and when the concentration level of NAD is lower than 10^{-4} M. The experimental results for τ_C/τ_0 listed in Table 2.21 well agree with the theoretical expectation.

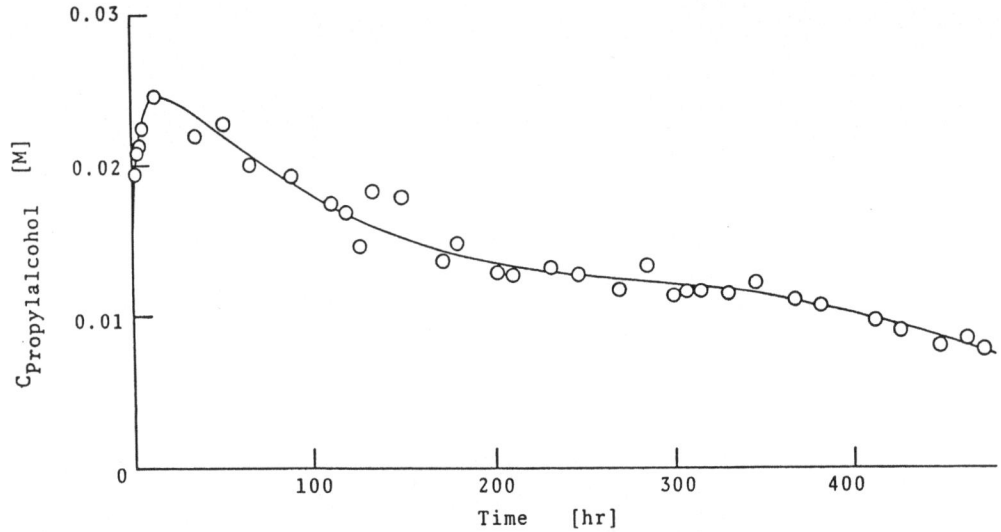

Fig. 2.25. Activity of affinity chromatographic reactor with a single enzyme (ADH) immobilized. ($C_{ADH}=4.15\times10^{-4}\,M$, $C_{NAD}=1\times10^{-4}\,M$, $\tau_0=2.01$ h, $\Delta t_{NAD}^{pulse}=30$ min, $C_{Propionaldehyde}=3.47\times10^{-2}\,M$, $C_{Ethanol}=0.835$ M)

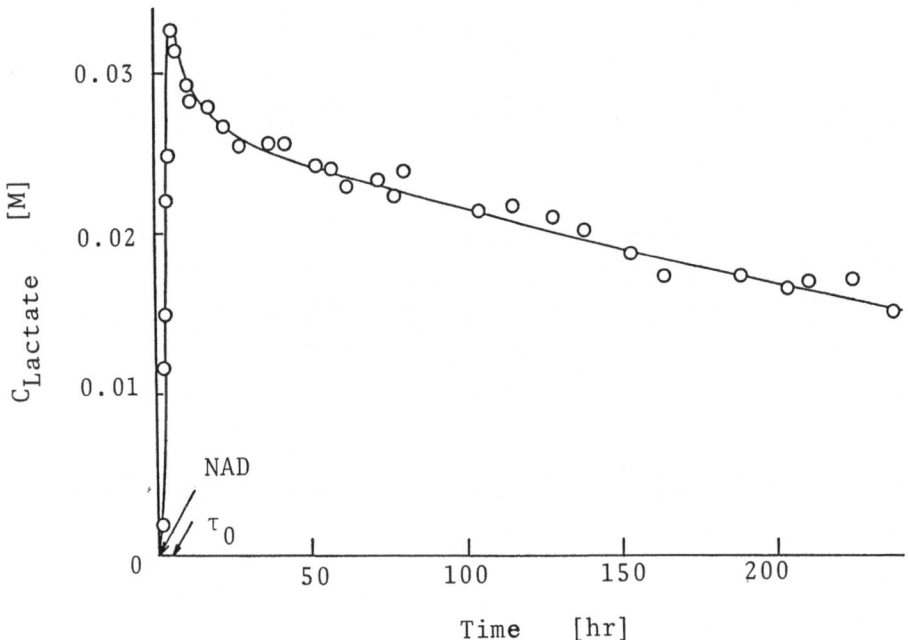

Fig. 2.26. Activity of affinity chromtographic reactor with conjugated enzymes (LDH and ADH) immobilized. ($C_{LDH}=4.68\times10^{-5}\,M$, $C_{ADH}=1.48\times10^{-4}\,M$, $C_{NAD}=5.10^{-5}\,M$, $\tau_0=4.24$ h, $\Delta t_{NAD}^{pulse}=20$ min, $C_{Pyruvate}$"$0.05\,M$, $C_{Ethanol}=0.835$ M)

150

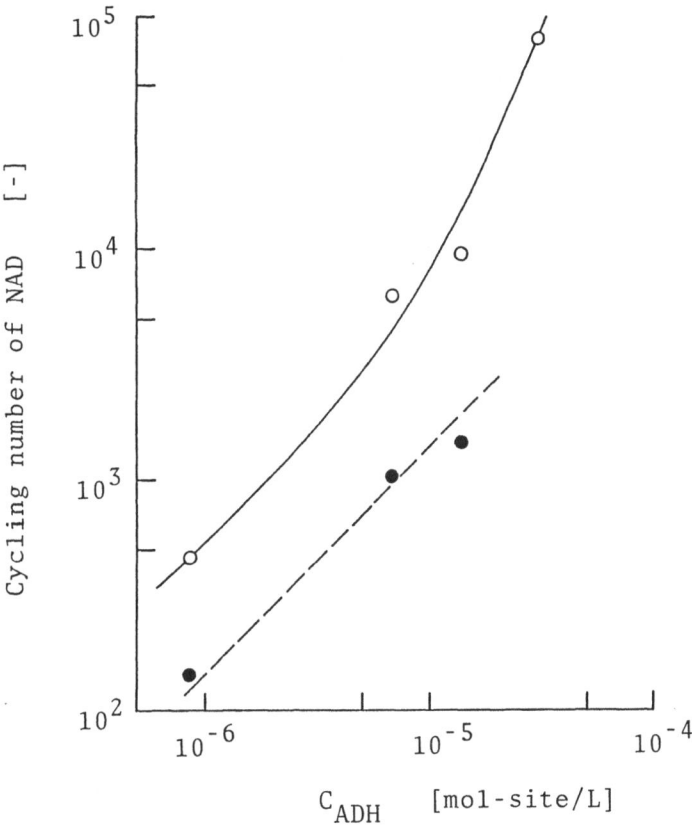

Fig. 2.27. Effect of immobilized enzyme concentration on cycling number of NAD. (C_{ADH}/$C_{LDH}=1.63$, $\tau_0=3.63$ h, $\Delta t_{NAD}^{pulse}=20$ min; —o—, $C_{NAD}=5\times10^{-5}$ M, – -•- –, $C_{NAD}=5\times10^{-4}$ M)

A typical performance of the affinity chromatographic reactor is shown in Fig. 2.25. In this case, NAD was fed as a pulse at 1×10^{-4} M for 30 min only at the beginning. After that period, only the substrates, propionaldehyde and ethanol, were fed continuously. The reactor operated for more than 400 h in spite of no additional feed of NAD. This fact clearly shows the expression of the "dynamic immobilizion" of NAD through affinity. The total cycling number of NAD, R_N, estimated from the half-life of the reactor activity was as high as 140 000.

The same principle was applied to the more general case with the conjugation of two enzymes [147]. A system with immobilized LDH and ADH in the ultrafiltration hollow-fiber capillary reactor was employed for this purpose. A typical result is shown in Fig. 2.26. NAD at $5+10^{-5}$ M was fed for 20 min only at the beginning. Thereafter, the substrates (pyruvate and ethanol) only were fed continuously. The reactor produced lactate from pyruvate with NADH cycling for more than 200 h without further feed of NAD. The cycling number of NAD, R_N, in this case was 412 000. Effects of the concentrations of the enzymes and the

151

coenzyme on R_N were tested; the results are shown in Fig. 2.27. The higher enzyme concentration and/or the lower NAD concentration gave the higher R_N, as was expected from the theoretical consideration for the simplified case described before by Eq. (2.54).

From the discussions given above, the operating mode of the bioreactor with continuous coenzyme cycling may be divided into three groups:

(1) Self-contained case with immobilized coenzyme
(2) Partially-contained case with no effect of the affinity between enzyme and coenzyme
(3) Partially-contained case with dynamic immobilization of coenzyme through the affinity

Comparisons among these modes should be done on a cost basis. The cost of catalysts in the self-contained case in US\$/mol-product, C^{SC}, may be expressed as following:

$$C^{SC} = s_C/(P_C \tau_C) + s_{E1}/(P_{E1} \tau_{E1}) + s_{E2}/(P_{E2} \tau_{E2}), \tag{2.55}$$

where s is the price of catalyst in US\$/mol for coenzyme and in US\$/U for enzymes. P is the productivity and τ is the average life-time of catalysts in days. Subscripts C, E1, and E2 correspond to coenzyme and enzymes for the main reaction and the regeneration reaction of coenzyme, respectively. Productivities of enzymes, P_{E1} and P_{E2}, are expressed in mol-product/day/U. These should be 0.00144 ($= 1\ \mu mol/min/U$) if the reactor is optimally operated. A lower value of the productivity than this reflects the existence of some depressing effects on enzyme reactions such as inhibition, inactivation of enzymes, mass transfer effects and so on. Productivity of coenzyme, P_C, is in another words the cycling rate and is expressed in mol-product/mol-coenyzme/day.

The cost of catalysts in the partially-contained case, C^{PC}, is expressed as following:

$$C^{PC} = s_C/R_N + s_{E1}/(P_{E1} \tau_{E1}) + s_{E2}/(P_{E2} \tau_{E2}), \tag{2.56}$$

where R_N is the cycling number of coenzyme.

Apparently, the difference of Eq. (2.56) from Eq. (2.55) exists only in the first term on the right side. However, there are substantial differences in each term in Eqs. (2.55) and (2.56) among the three modes mentioned before.

In mode (1), the price of coenzyme, s_C, is high because of the need for sophisticated chemical modification. In addition, the immobilized coenzyme generally has a limited life-time. Therefore, the coenzyme cycling number should be expressed as the cycling rate (P_C) multiplied by the life-time of the immobilized coenzyme (τ_C). As for polyethylene glycol-bound NAD, HAYAKAWA et al. [177] and OHSHIMA et al. [176] reported 5.7 days and 15 days for the half-lives, respectively. The cycling rate of coenzyme (P_C) in this mode is generally high [174, 176]. The productivities of enzymes, P_{E1} and P_{E2}, are from moderate to high.

In mode (2), the price of coenzyme, s_C, is that for the native one. The cycling number, R_N, and the productivities of the enzymes, P_{E1} and P_{E2}, are from low to medium. This mode has an advantage of easy application because of no involvement of any sophisticated technique as in the chemical modification of coenzyme in mode (1).

In mode (3), the price of coenzyme is also that for the native one. This mode requires, however, high concentrations of immobilized enzymes, which lowers the productivities per enzyme (P_{E1}, P_{E2}). The cycling number is extremely high especially in the case of the affinity chromatographic reactor.

2.6 Concluding Remarks

The discussions given in this chapter clearly show many possibilities of the bioreactor with coenzyme cycling for chemical transformations, synthesis, and analytical purposes as shown in Tables from 2.6 to 2.12. Nevertheless, it has not been well used, mainly because coenzymes involved are still believed to be expensive and effective methods of coenzyme regeneration have been lacking. However, a wide choice of methods for coenzyme regeneration is available now, as described in Sect. 2.2, and some of them seem practical. Various processes involving coenzyme cycling have been investigated so far as shown in Sect. 2.3. Coenzyme cycling in analytical use is another field of interest because coenzymes often play key roles in enzyme assays.

In a continuous operation, the methods with chemically modified coenzyme (Sect. 2.4) and the techniques utilizing the affinity interaction between enzyme and coenzyme (Sect. 2.5) make it possible to achieve a cycling number of coenzyme from 10 000 to 400 000. Therefore, the cost of coenzymes can be considered well reduced now. The next problem will be to accumulate more practical data on the stabilities of enzymes involved and coenzyme (in the case of chemically modified coenzyme) and on the reactor kinetics at a high substrate concentration for the reduction of the cost in product purification. Then, cost estimations for a specific process would become possible and a practical application of the enzyme process with coenzyme cycling could be considered. It should be mentioned that enzyme processes as described here are always competing with the organic process and fermentation process. Advantages of the enzyme process exist in its rapid kinetics, high selectivity, and stereospecificity.

2.7 Abbreviations

AcK:	Acetate kinase
AdK:	Adenylate kinase
ADH:	Alcohol dehydrogenase
sec ADH:	Secondary alcohol dehydrogenase
ADP:	Adenosine 5'-diphosphate
AE-NAD:	Aminoethyl-NAD
AlaDH:	Alanine dehydrogenase
ALDDH:	Aldehyde dehydrogenase
AMP:	Adenosine 5'-monophosphate

ATP: Adenosine 5'-triphosphate
ATP-γ-S: Adenosine 5'-O-(3-thiotriphosphate)
CK: Carbamate kinase
CoA: Coenzyme A
CTP: Cytidine 5'-triphosphate
DH: Dehydrogenase
DIA: Diaphorase
FAD: Flavin adenine dinucleotide
FDH: Formate dehydrogenase
FMN: Flavin mononucleotide
GDH: Glucose dehydrogenase
GlutDH: Glutamate dehydrogenase
GTP: Guanosine 5'-triphosphate
G6P: Glucose 6-phosphate
G6PDH: Glucose 6-phosphate dehydrogenase
G6S: Glucose 6-sulfate
HK: Hexokinase
HLADH: Horse liver alcohol dehydrogenase
LDH: Lactate dehydrogenase
MDH: Malate dehydrogenase
MV: Methyl viologen
NAD: Nicotinamide adenine dinucleotide
NADP: Nicotinamide adenine dinucleotide phosphate
NTP: Nucleotide triphosphate
PAPS: 3'-Phosphoadenosine-5'-phosphosulfate
PEG: Polyethylene glycol
PEG–NAD: Polyethylene glycol-bound NAD
PEP: Phosphoenolpyruvate
6-PG: 6-Phosphogluconate
6-PGDH: 6-Phosphogluconate dehydrogenase
PK: Pyruvate kinase
PMS: Phenazine methosulfate
PRI: Phosphoriboisomarase
SAM: S-Adenosylmethionine
TADH: Thermostable alcohol dehydrogenase
T-LDH: Thermostable lactate dehydrogenase
T-MDH: Thermostable malate dehydrogenase
UTP: Uridine 5'-triphosphate
YADH: Yeast alcohol dehydrogenase

References

1. Wang SS, King CK (1979) The use of coenzymes in biochemical reactors. In: Ghose TK (Ed) Advances in biochemical engineering. Springer, Berlin Heidelberg New York, p 119
2. Martinek K, Semenov AN (1981) J Appl Biochem 3:93
3. Sharma BP, Bailey LF, Messing RA (1982) Angew Chem Int Ed Engl 21:837
4. Lowe CR (1983) Phil Trans Roy Soc London B300:335

154

5. Mosbach K (1983) Phil Trans Roy Soc London B300:355
6. Vandamme EJ (1983) Enzyme Microb Technol 5:403
7. May SW, Padgette SR (1983) Bio/Technology, p 677
8. Bowen R, Pugh S (1985) Chem Ind 20:323
9. Whitesides GM, Wong CH (1985) Angew Chem Int Ed Engl 24:617
10. Barman TE (1969) Enzyme handbook. Springer, Berlin Heidelberg New York
11. Chibata I, Tosa T (1981) In: Fukui S, Chibata I, Suzuki S (eds) Koso-Kogaku. Tokyo Kagaku Dojin, Tokyo, p 373
12. Cilento G (1960) Arch Biochem Biophys 88:352
13. Spiegel MJ, Drydale GR (1960) J Biol Chem 235:2498
14. Drydale GR, Spiegel MJ, Strittmatter (1961) J Biol Chem 236:2323
15. Ludowieg J, Levy A (1964) Biochemistry 3:373
16. Abril O, Whitesides GM (1982) J Am Chem Soc 104:1552
17. Wienkamp R, Steckhan E (1983) Angew Chem Ind Ed Engl 22:497
18. Aizawa M, Coughlin RW, Charles M (1976) Biochim Biophys Acta 440:233
19. Aizawa M, Coughlin RW, Charles M (1976) Biotech Bioeng 18:209
20. Aizawa M, Suzuki S, Kubo M (1976) Biochim Biophys Acta 444:886
21. Aizawa M, Ikariyama Y, Suzuki S (1976) J Solid-Phase Biochem 1:249
22. Hatanaka A, Adachi O, Chiyonobu T, Amemiya M (1971) Agric Biol Chem 35:1304
23. Vandecasteele JP (1980) Appl Environ Microbiol 39:327
24. Wong CH, Whitesides GM (1982) 47:2816
25. Theorell H, McKinley-McKee JS (1961) Acta Chem Scand 15:1797
26. Sund H, Theorell H (1962) Alcohol dehydrogenases. In: Boyer, Lardy, Myrback (eds) The enzymes, 2nd edn, vol 7. Academic Press, New York, p 25
27. Gupta NK, Robinson WG (1966) Biochim Biophys Acta 118:431
28. Dickenson CJ, Dickinson FM (1975) Biochem J 147:303
29. Yamazaki Y, Maeda H, Suzuki H (1976) Biotech Bioeng 18:1761
30. Lamed RJ, Zeikus JG (1980) J Bacteriol 141:1251
31. Lamed RJ, Zeikus JG (1981) Biochem J 195:183
32. Godbole SS, D'Souza SF, Nadkarni GB (1983) Enzyme Microb Tech 5:125
33. Sugawara Y, Sasaki S (1977) Biochim Biophys Acta 480:343
34. Tamaki N, Nakamura M, Kimura K, Hama T (1977) J Biochem 82:73
35. Ramaley RF, Vasantha N (1983) J Biol Chem 258:12558
36. Okuno H, Nagata K, Nakajima H (1985) J Appl Biochem 7:192
37. Noltmann EA, Gubler CJ, Kuby SA (1961) J Biol Chem 236:1225
38. Wong CH, Gordon J, Cooney CL, Whitesides GM (1981) J Org Chem 46:4676
39. Wong CH, Whitesides GM (1981) J Am Chem Soc 103:4890
40. Izumi Y, Mishra SK, Ghosh BS, Tani Y, Yamada H (1983) J Ferment Technol 61:135
41. Shutte H, Flossdorf J, Sahm H, Kula MR (1976) Eur J Biochem 62:151
42. Shaked Z, Whitesides GM (1980) J Am Chem Soc 102:7104
43. Hu ASL, Cline AL (1964) Biochim Biophys Acta 93:237
44. Shneider K, Schlegel HG (1976) Biochim Biophys Acta 452:66
45. Zeikus JG, Fuchs G, Kenealy W, Thauer RK (1977) J Bacteriol 132:604
46. Krasna AI (1979) Enzyme Microb Technol 1:165
47. Ashton WT, Brown RD, Jacobson F, Walsh C (1979) J Am Chem Soc 101:4419
48. Gunsalus RP, Wolfe RS (1980) J Biol Chem 255:1891
49. Wong CH, Daniels L, Orme-Johnson WH, Whitesides GM (1981) J Am Chem Soc 103:6227
50. Massey V (1958) Biochim Biophys Acta 30:205
51. Massey V (1960) Biochim Biophys Acta 37:314
52. Massey V, Veeger C (1961) Biochim Biophys Acta 48:33
53. Kaplan F, Setlow P, Kaplan NO (1969) Arch Biochem Biophys 132:91
54. Scouten WH, Knowles H, Freitag LC, Iobst W (1977) Biochim Biophys Acta 482:11
55. Sakurai Y, Fukuyoshi Y, Hamada M, Hayakawa T, Koike M (1970) J Biol Chem 245:4453
56. Avigad G, Alroy Y, England S (1968) J Biol Chem 243:1936
57. Karube I, Otsuka T, Kayano H, Matsunaga T, Suzuki S (1980) Biotech Bioeng 22:2655

58. Eguchi SY, Nishio N, Nagai S (1983) Agric Biol Chem 47:2941
59. Eguchi SY, Nakata H, Nishio N, Nagai S (1984) Appl Microbiol Biotech 20:213
60. Suye S, Yokoyama S (1985) Enzyme Microb Tech 7:418
61. Lehninger AL (1975) Biochemistry, 2nd edn. Worth Publishers, New York, pp 316, 340
62. Jones JB, Sneddon DW, Higgins W, Lewis AJ (1972) JCS Chem Comm, p 856
63. Jones JB, Taylor KE (1973) JCS Chem Comm, p 205
64. Chambers RP, Ford JR, Allender JH, Baricos WH, Cohen W (1974) Continuous processing with cofactor requiring enzymes: coenzyme retention and regeneration. In: Pye EK, Wingard LB (eds) Enzyme engineering, vol 2. Plenum Press, New York, p 195
65. Wagner F, Convit J, Bernt E, Nelbock M (1964) Angew Chem Internat Edit 3:587
66. Mansson MO, Mattiasson B, Gestrelius S, Mosbach K (1976) Biotech Bioeng 18:1145
67. Jones JB, Taylor KE (1976) Can J Chem 54:2969
68. Taylor KE, Jones JB (1976) J Am Chem Soc 98:5689
69. Jones JB, Taylor KE (1976) Can J Chem 54:2974
70. Sober HA (ed) (1970) Handbook of biochemistry, 2nd edn. The Chemical Rubber Co, Cleveland, Ohio, J-33
71. Sigma catalogue (1986) Sigma Chemical Co., St. Louis, MO
72. Aizawa M, Coughlin RW, Charles M (1975) Biochim Biophys Acta 385:362
73. Aizawa M, Ikariyama Y, Suzuki S (1976) J Solid-Phase Biochem 1:197
74. Kelly RM, Kirwan DJ (1977) Biotech Bioeng 19:1215
75. Jaegfeldt H, Torstensson A, Johansson G (1978) Anal Chim Acta 97:221
76. Moiroux J, Elving PJ (1979) Anal Chem 51:346
77. Huck H, Schmidt HL (1981) Angew Chem Int Ed Engl 20:402
78. Jaegfeldt H, Torstensson ABC, Gorton LGO, Johansson G (1981) Anal Chem 53:1979
79. Torstensson A, Johansson G, Mansson MO, Larsson PO, Mosbach K (1980) Anal Lett 13(B10):837
80. Fahien LA, Wiggert BO, Cohen PP (1965) J Biol Chem 240:1083
81. Tomkins GM, Yielding KL, Curran JF, Summers MR, Bitensky MW (1965) J Biol Chem 240:3793
82. Hooper AB, Terry KR, Kemp KD (1974) Biochim Biophys Acta 358:14
83. Schwert GW, Winer AD (1963) Lactate dehydrogenase. In: Boyer PD, Lardy H, Myrback K (eds) The enzymes, 2nd edn, vol 7. Academic Press, New York, p 127
84. Moiroux J, Elving PJ (1978) Anal Chem 50:1056
85. Jaegfeldt H (1980) J Electroanal Chem 110:295
86. Watanabe H, Hastings JW (1982) Molecul Cell Biochem 44:181
87. Hoskins DD, Whiteley HR, Mackler B (1962) J Biol Chem 237:2647
88. Kawai K, Eguchi Y (1975) J Ferment Tech 53:588
89. Gwak SH, Ota Y, Yagi O, Minoda Y (1982) J Ferment Tech 60:205
90. Asada Y, Miyabe M, Kikkawa M, Kuwahara M (1986) Agric Biol Chem 50:525
91. Dolin MI (1957) J Biol Chem 225:557
92. Burstein C, Ounissi H, Legoy MD, Gellf G, Thomas D (1981) Appl Biochem Biotech 6:329
93. Estival F, Burstein C (1985) Enz Microb Tech 7:29
94. Langer RS, Hamilton BK, Gardner CR, Archer MC, Colton CK (1976) AIChE J 22:1079
95. Whitesides GM, Siegel M, Garrett P (1975) J Org Chem 40:2516
96. Crans DC, Whitesides GM (1983) J Org Chem 48:3130
97. Nakajima H, Suzuki K, Imahori K (1978) J Biochem 84:193
98. Nakajima H, Nagata K, Kondo H, Imahori K (1984) J Appl Biochem 6:19
99. Kondo H, Tomioka I, Nakajima H, Imahori K (1984) J Appl Biochem 6:29
100. Whitesides GM, Lamotte AL, Adalsteinsson O, Baddour RF, Chmurny AC, Colton CK, Pollak A (1979) J Mol Cat 6:177
101. Adalsteinsson O, Lamotte A, Baddour RF, Colton CK, Pollak A, Whitesides GM (1979) J Mol Cat 6:199
102. Tanaka A, Hagi N, Gellf G, Fukui S (1980) Agric Biol Chem 44:2399
103. Grisolia S, Harmon P, Raijman L (1962) Biochim Biophys Acta 62:293
104. Thorne KJI, Jones ME (1963) J Biol Chem 238:2992

105. Kalman SM, Duffield PH (1964) Biochim Biophys Acta 92:498
106. Marshall DL (1973) Biotech Bioeng 15:447
107. Manca de Nadra MC, Nadra Chaud CA, Pesce de Ruiz Holgado A, Oliver G (1986) Biotech Appl Biochem 8:46
108. Tietz A, Ochoa S (1958) Arch Biochem Biophys 78:477
109. Hirschbein BL, Mazenod FP, Whitesides GM (1982) J Org Chem 47:3765
110. Slegers G, De Laet S, Lambrecht RH, Block C (1986) Enz Microb Tech 8:153
111. Kuby SA, Noda L, Lardy HA (1954) J Biol Chem 210:65
112. Sakata K, Kitano H, Ise N (1981) J Appl Biochem 3:518
113. Asada M, Nakanishi K, Matsuno R, Kariya Y, Kimura A, Kamikubo T (1978) Agric Biol Chem 42:1533
114. Asada M, Morimoto K, Nakanishi K, Matsuno R, Tanaka A, Kimura A, Kamikubo T (1979) Agric Biol Chem 43:1773
115. Asada M, Shirai Y, Nakanishi K, Matsuno R, Kimura A, Kamikubo T (1981) J Ferment Tech 59:239
116. Asada M, Yamamoto K, Nakanishi K, Matsuno R, Kimura A, Kamikubo T (1981) Eur J Appl Microbiol Biotech 12:198
117. Fujio T, Furuya A (1983) J Ferment Tech 61:261
118. Fukuda Y, Yamaguchi S, Hashimoto H, Shimosaka M, Kimura A (1984) Agric Biol Chem 48:2877
119. Matsuoka H, Suzuki S (1981) Biotech Bioeng 23:1103
120. Tani Y, Mitani Y, Yamada H (1984) Agric Biol Chem 48:431
121. Sawa Y, Kanayama K, Ochiai H (1982) Biotech Bioeng 24:305
122. Smeds AL, Veide A, Enfors SO (1983) Enz Microb Tech 5:23
123. Garde VL, Cocquempot MF, Barbotin JN, Thomasset B, Thomas D (1980) Immobilized thylakoids and chromatophores: hydrogen production and ATP regeneration. In: Weetall HH, Royer GP (eds) Enzyme engineering, vol 5. Plenum Press, New York, p 109
124. Cocquempot MF, Larreta Garde V, Thomas D (1980) Biochimie 62:615
125. Larreta Garde V, Gellf G, Thomas D (1981) Eur J Biochem 116:337
126. Larreta Garde V, Thomasset B, Tanaka A, Gellf G, Thomas D (1981) Eur J Appl Microbiol Biotech 11:133
127. Larreta Garde V, Gellf G, Thomas D (1982) Eur J Appl Microb Biotech 14:232
128. Shaked Z, Barber JJ, Whitesides GM (1981) J Org Chem 46:4100
129. DiCosmo R, Wong CH, Daniels L, Whitesides GM (1981) J Org Chem 46:4622
130. Abril O, Whitesides GM (1982) J Am Chem Soc 104:1552
131. Wienkamp R, Steckhan E (1982) Angew Chem Int Ed Engl 21:782
132. Mandler D, Willner I (1984) J Am Chem Soc 106:5352
133. Maeda H, Kajiwara S (1985) Biotech Bioeng 27:596
134. Langer RS, Gardner CR, Hamilton BK, Colton CK (1977) AIChE J 23:1
135. Fink DJ, Rodwell VW (1975) Biotech Bioeng 17:1029
136. Davies J, Jones JB (1979) J Am Chem Soc 101:5405
137. Lamed RJ, Keinan E, Zeikus JG (1981) Enz Microb Tech 3:144
138. Dodds DR, Jones JB (1982) J Chem Soc Chem Comm, p 1080
139. Nakarai M, Chikamatsu H, Taniguchi M (1982) Chem Lett, p 1761
140. Sugai T, Fujita M, Mori K (1983) Nippon Kagaku Kaishi, p 1315
141. Miyawaki O, Nakamura K, Yano T (1986) Proceeding of World Congress 3 of Chemical Engining, vol 1. Tokyo, September, p 965
142. Levy HR, Loewus FA, Vennesland B (1957) J Am Chem Soc 79:2949
143. Davies P, Mosbach K (1974) Biochim Biophys Acta 370:329
144. Wykes JR, Dannill P, Lilly MD (1975) Biotech Bioeng 17:51
145. Morikawa Y, Karube I, Suzuki S (1978) Biochim Biophys Acta 523:263
146. Miyawaki O, Nakamura K, Yano Y (1982) J Chem Eng Japan 15:224
147. Miyawaki O, Nakamura K, Yano Y (1985) Agric Biol Chem 49:2063
148. Wingard LB, Roach RP, Miyawaki O, Egler KA, Klinzing GE, Silver RS, Brackin JS (1985) Enz Microb Tech 7:503
149. Miyawaki O, Wingard LB, Brackin JS, Silver RS (1986) Biotech Bioeng 28:343
150. Hou CT, Patel RN, Laskin AI, Barnabe N (1982) J Appl Biochem 4:379
151. Patel RN, Hou CT, Laskin AI, Felix A (1982) J Appl Biochem 4:175

152. Righini-Tapie A, Azerad R (1984) J Appl Biochem 6:361
153. Kitpreechavanich V, Nishio N, Hayashi M, Nagai S (1985) Biotech Lett 7:657
154. Carrea G, Bovara R, Longhi R, Barani R (1984) Enz Microb Tech 6:307
155. Matsunaga T, Matsunaga N, Nishimura S (1985) Biotech Bioeng 27:1277
156. Campbell J, Chang TMS (1976) Biochem Biophys Res Comm 69:562
157. Grunwald J, Chang TMS (1978) Biochem Biophys Res Comm 81:565
158. Grunwald J, Chang TMS (1979) J Appl Biochem 1:104
159. Chang TMS (1980) New approaches using immobilized enzymes for the removal of urea and ammonia. In: Weetall HH, Royer GP (eds) Enzyme engineering, vol 5. Plenum Press, New York, p 225
160. Yu YT, Chang TMS (1981) FEBS Lett 125:94
161. Yu YT, Chang TMS (1982) Enz Microb Tech 4:327
162. Chang TMS, Yu YT, Grunwald J (1982) Artificial cell immobilized multienzyme systems and cofactors. In: Chibata I, Fukui S, Wingard LB (eds) Enzyme engineering, vol 6. Plenum Press, New York, p 451
163. Passingham BJ, Barton RN (1975) Anal Biochem 65:418
164. Maurer PJ, Miller MJ (1983) J Am Chem Soc 105:240
165. Wong CH, Gordon J, Cooney CL, Whitesides GM (1981) J Org Chem 46:4676
166. Wong CH, Whitesides GM (1981) J Am Chem Soc 103:4890
167. Wong CH, Whitesides GM (1982) J Am Chem Soc 104:3542
168. Hirschbein BL, Whitesides GM (1982) J Am Chem Soc 104:4458
169. Wong CH, Whitesides GM (1983) J Am Chem Soc 105:5012
170. Wong CH, Drueckhammer DG, Sweers HM (1985) J Am Chem Soc 107:4028
171. Wichmann R, Wandrey C (1981) Biotech Bioeng 23:2789
172. Wandrey C, Wichmann R (1980) Immobilization of biocatalysts using ultrafiltration techniques. In: Weetall HH, Royer GP (eds) Enzyme engineering, vol 5. Plenum Press, New York, p 453
173. Wandrey C, Wichmann R, Jandel AS (1982) Multi enzyme systems in membrane reactors. In: Chibata I, Fukui S, Wingard LB (eds) Enzyme engineering, vol 6. Plenum Press, New York, p 61
174. Wichmann R, Wandrey C, Hummel W, Schutte H, Buckmann AF, Kula MR (1984) Ann NY Acad Soc 434:87
175. Wandrey C, Fiolitakis E, Wichmann U, Kula MR (1984) Ann NY Acad Sci 434:90
176. Ohshima T, Wandrey C, Kula MR, Soda K (1985) Biotech Bioeng 27:1616
177. Hayakawa K, Urabe I, Okada H (1985) J Ferment Tech 63:245
178. Tischer W, Bader J, Simon H (1979) Eur J Biochem 97:103
179. Tischer W, Tiemeyer W, Simon H (1980) Biochimie 62:331
180. Bader J, Simon H (1980) Arch Microbiol 127:279
181. Simon H, Gunther H, Bader J, Tischer W (1981) Angew Chem Int Ed Engl 20:861
182. Giesel H, Simon H (1983) Arch Microbiol 135:51
183. Bader J, Gunther H, Nagata S, Schuetz HJ, Link ML, Simon H (1984) J Biotech 1:95
184. Simon H, Bader J, Gunther H, Neumann S, Thanos J (1984) Ann NY Acad Sci 434:171
185. Simon H, Bader J, Gunther H, Neumann S, Thanos J (1985) Angew Chem Int Ed Engl 24:539
186. Irwin AJ, Jones JB (1977) J Am Chem Soc 99:1625
187. Irwin AJ, Jones JD (1977) J Am Chem Soc 99:556
188. Jakovak IJ, Ng G, Lok KP, Jones JB (1980) JCS Chem Comm, p 515
189. Jones JB, Finch MAW, Jakovak IJ (1982) Can J Chem 60:2007
190. Jakovak IJ, Goodbrand HB, Lok KP, Jones JB (1982) J Am Chem Soc 104:4659
191. Bridges AJ, Raman PS, Ng GSY, Jones JB (1984) J Am Chem Soc 106:1461
192. Cremonesi P, Carrea G, Ferrara L, Antonini E (1974) Eur J Biochem 44:401
193. Legoy MD, Larreta Garde V, Ergan F, Thomas D (1979) J Solid-Phase Biochem 4:143
194. Carrea G, Bovara R, Longhi R, Riva S (1985) Enz Microb Tech 7:597
195. Legoy MD, Kim HS, Thomas D (1985) Process Biochem, October, p 145
196. Chambers RP, Walle EM, Baricos WH, Cohen W (1978) High turnover NAD regeneration in the coupled dehydrogenase conversion of sorbitol to fructose. In: Pye EK, Weetall HH (eds) Enzyme engineering, vol 3. Plenum Press, New York, p 363

158

197. Ergan F, Thomas D, Chang TMS (1984) Appl Biochem Biotech 10:61
198. Laval JM, Bourdillon C, Moiroux J (1984) J Am Chem Soc 106:4701
199. Coughlin RW, Aizawa M, Alexander BF, Charles M (1975) Biotech Bioeng 17:515
200. Coughlin RW, Alexander BF (1975) Biotech Bioeng 17:1379
201. May SW, Landgraff LM (1976) Biochem Biophys Res Comm 68:786
202. Burstein C, Ounissi H, Legoy MD, Gellf G, Thomas D (1981) Appl Biochem Biotech 6:329
203. Chave E, Adamowicz E, Burstein C (1982) Appl Biochem Biotech 7:431
204. Campbell J, Chang TMS (1975) Biochim Biophys Acta 397:101
205. Chang TMS, Kuntarian N (1978) Galactose conversion using a microcapsule immobilized multienzyme cofactor recycling system. In: Broun GB, Manecke G, Wingard LB (eds) Enzyme engineering, vol 4. Plenum Press, New York, p 193
206. Murata K, Tani K, Kato J, Chibata I (1978) Eur J Appl Microbiol Biotech 6:23
207. Murata K, Tani K, Kato J, Chibata I (1979) J Appl Biochem 1:283
208. Murata K, Tani K, Kato J, Chibata I (1980) Eur J Appl Microbiol Biotech 10:11
209. Murata K, Tani K, Kato J, Chibata I (1980) Biochimie 62:347
210. Murata K, Tani K, Kato J, Chibata I (1981) Eur J Appl Microbiol Biotech 11:72
211. Murata K, Kimura A (1981) Tanpakushitsu-Kakusan-Koso 26:915
212. Murata K, Tani K, Kato J, Chibata I (1981) Enz Microb Tech 3:233
213. Miyawaki O, Nakamura K, Yano T (1982) Agric Biol Chem 46:2725
214. Walt DR, Findeis MA, Rios-Mercadillo VM, Auge J, Whitesides GM (1984) J Am Chem Soc 106:234
215. Kimura A, Tatsutomi Y, Mizushima N, Tanaka A, Matsuno R, Fukuda H (1978) Eur J Appl Microbiol Biotech 5:13
216. Kimura A, Tatsutomi Y, Matsuno R, Tanaka A, Fukuda H (1981) Eur J Appl Microbiol Biotech 11:78
217. Ado Y, Suzuki Y, Tadokoro T, Kimura K, Samejima H (1979) J Solid-Phase Biochem 4:43
218. Samejima H, Kimura K, Ado Y, Suzuki Y, Tadokoro T (1978) Regeneration of ATP by immobilized microbial cells and its utilization for the synthesis of nucleosides. In: Broun GB, Manecke G, Wingard LB (eds) Enzyme engineering, vol 4. Plenum Press, New York, p 237
219. Watanabe S, Shirota S, Haneda K, Takeda I (1981) J Ferment Tech 59:191
220. Watanabe S, Kitajima N, Shirota S, Takeda I (1981) J Ferment Tech 59:197
221. Shih YS, Whitesides GM (1977) J Org Chem 42:4165
222. Pollak A, Baughn RL, Whitesides GM (1977) J Am Chem Soc 99:2366
223. Rios-Mercadillo VM, Whitesides GM (1979) J Am Chem Soc 101:5828
224. England R, Eisenthal R, Gacesa P, Whish W (1979) Biochem Soc Trans 7:13
225. Berke W, Morr M, Wandrey C, Kula MR (1984) Ann NY Acad Sci 434:257
226. Stramondo JG, Solomon BA, Colton CK, Wang DIC (1976/77) AIChE Symp Ser 74:1
227. Siegbahn N, Mosbach K, Grodzki K, Zocher R, Madry N, Kleinkauf H (1985) Biotech Lett 7:297
228. Nakajima H, Kitabatake S, Tsurutani R, Tomioka I, Yamamoto K, Imahori K (1984) Biochim Biophys Acta 790:197
229. Ikariyama Y, Aizawa M, Suzuki S (1979) J Solid-Phase Biochem 4:69
230. Ikariyama Y, Aizawa M, Suzuki S (1979) J Solid-Phase Biochem 4:279
231. Gesrelius S, Mansson MO, Mosbach K (1975) Eur J Biochem 57:529
232. Venn RF, Larsson PO, Mosbach K (1977) Acta Chem Scand B31:141
233. Mansson MO, Larsson PO, Mosbach K (1978) Eur J Biochem 86:455
234. Mansson MO, Larsson PO, Mosbach K (1979) FEBS Lett 98:309
235. Yamazaki Y, Maeda H (1982) Agric Biol Chem 46:1571
236. Yamazaki Y, Maeda H, Kamibayashi A (1982) Biotech Bioeng 24:1915
237. Legoy MD, Le Moullec JM, Thomas D (1978) FEBS Lett 94:335
238. Legoy MD, Larreta Garde V, Le Moullec JM, Ergan F, Thomas D (1980) Biochimie 62:341
239. Mazid MA, Laidler KJ (1982) Biotech Bioeng 24:2087
240. Wong CH, McCurry SD, Whitesides GM (1980) J Am Chem Soc 102:7938

241. Abril O, Crans DC, Whitesides GM (1984) J Org Chem 49:1360
242. Mosbach K, Larsson PO, Lowe C (1976) Immobilized coenzymes. In: Mosbach K (ed) Methods in enzymology, vol 44. Academic Press, New York, p 859
243. Yamazaki Y, Maeda H, Suzuki H (1977) Hakko-To-Kogyo 35:270
244. Okada H (1978) Hakko-Kogaku 56:441
245. Okada H, Urabe I (1983) Hakko-To-Kogyo 41:172
246. Yamazaki Y, Maeda H (1983) Yukigosei-Kagaku 41:1088
247. Maeda H (1984) Enzyme Engineering News No 11, p 12
248. Maeda H (1985) Saikin-No-Kagaku-Kogaku vol 37, The Society of Chemical Engineers, Japan, p 118
249. Urabe I (1986) Hakko-Kogaku 64:77
250. Urabe I (1981) Hokoso-no-koteika. In: Fukui S, Chibata I, Suzuki S (eds) Koso Kogaku. Tokyo Kagaku Dojin, Tokyo, p 203
251. Chandrasekhar K, McPherson A, Adams MJ, Rossmann MG (1973) J Mol Biol 76:503
252. Adams MJ, Buehner M, Chandrasekhar K, Ford GC, Hackert ML, Liljas A, Rossmann MG, Smiley IE, Allison WS, Everse J, Kaplan NO, Taylor SS (1973) Proc Nat Acad Sci USA 70:1968
253. Jones JW, Robins RK (1963) J Am Chem Soc 85:193
254. Brink JJ, Schein AH (1963) J Med Pharm Chem 6:563
255. Falbriard JG, Posternak TH, Sutherland EW (1967) Biochim Biophys Acta 148:99
256. Mosbach K, Guilford H, Ohlsson R, Scott M (1972) Biochem J 127:625
257. Larsson PO, Mosbach K (1971) Biotech Bioeng 13:393
258. Lindberg M, Larsson PO, Mosbach K (1973) Eur J Biochem 40:187
259. Larsson PO, Mosbach K (1974) FEBS Lett 46:119
260. Wykes JR, Dunnill P, Lilly MD (1972) Biochim Biophys Acta 286:260
261. Zappelli P, Rossodivita A, Re L (1975) Eur J Biochem 54:475
262. Zappelli P, Rossodivita A, Prosperi G, Pappa R, Re L (1976) Eur J Biochem 62:211
263. Schmidt HL, Grenner G (1976) Eur J Biochem 67:295
264. Muramatsu M, Urabe I, Yamada Y, Okada H (1977) Eur J Biochem 80:111
265. Furukawa S, Katayama N, Iizuka T, Urabe I, Okada H (1980) FEBS Lett 121:239
266. Sakaguchi Y, Murachi T (1980) J Appl Biochem 2:117
267. Fuller CW, Rubin JR, Bright HJ (1980) Eur J Biochem 103:421
268. Buckmann AF, Kula MR, Wichmann R, Wandrey C (1981) J Appl Biochem 3:301
269. Yoshikawa M, Goto M, Ikura K, Sasaki R, Chiba H (1982) Agric Biol Chem 46:207
270. Adachi S, Ogata M, Tobita H, Hashimoto K (1984) Enz Microb Tech 6:259
271. Furukawa S, Sugimoto Y, Urabe I, Okada H (1980) Biochimie 62:629
272. Yamazaki Y, Maeda H (1981) Agric Biol Chem 45:2277
273. Mosbasch K, Larsson PO (1978) Immobilized cofactors and cofactor fragments in general ligand affinity chromatography and as active cofactors. In: Pye EK, Weetall HH (eds) Enzyme engineering, vol 3. Plenum Press, New York, p 291
274. Flygare S, Mannsson MO, Larsson PO, Mosbach K (1982) Appl Biochem Biotech 7:59
275. Mansson MO, Siegbahn N, Mosbach K (1983) Proc Natl Acad Sci USA 80:1487
276. Siegbahn N, Mansson MO, Mosbach K (1986) Appl Biochem Biotech 12:91
277. Katayama N, Hayakawa K, Urabe I, Okada H (1984) Enz Microb Tech 6:538
278. Eguchi T, Kanzaki N, Kagotani T, Taniguchi T, Urabe I, Okada H (1985) J Ferment Tech 63:563
279. Inoue T, Urabe I, Yamada Y, Okada H (1985) J Ferment Tech 63:485
280. Lowe CR, Mosbach K (1974) Eur J Biochem 49:511
281. Zappelli P, Pappa R, Rossodivita A, Re L (1977) Eur J Biochem 72:309
282. Urabe I, Okuda K, Okada H (1985) Kagaku-To-Seibutsu 23:58
283. Okuda K, Suntinanalerts P, Miyoshi S, Urabe I, Yamada Y, Okada H (1985) Eur J Biochem 147:241
284. Okuda K, Urabe I, Okada H (1985) Eur J Biochem 147:249
285. Okuda K, Urabe I, Okada H (1985) Eur J Biochem 151:33
286. Araki H, Yamazaki Y, Maeda H, Satoh A (1984) J Ferment Tech 62:221
287. Fuller CW, Bright HJ (1977) J Biol Chem 252:6631
288. Yamazaki Y, Maeda H, Suzuki H (1977) Eur J Biochem 77:511

289. Yamazaki Y, Suzuki H (1978) Eur J Biochem 92:197
290. Yamazaki Y, Maeda H (1981) Agric Biol Chem 45:2091
291. Zappelli P, Pappa R, Rossodivita A, Re L (1978) Eur J Biochem 89:491
292. Le Goffic F, Sicsic S, Vincent C (1978) Biochimie 60:421
293. Rieke E, Barry S, Mosbach K (1979) Eur J Biochem 100:203
294. Lowry OH, Passonneau JV, Schulz DW, Rock MK (1961) J Biol Chem 236:2746
295. Kato T, Berger SJ, Carter JA, Lowry OH (1973) Anal Biochem 53:86
296. Kato T, Lowry OH (1973) J Biol Chem 248:2044
297. Kato T (1975) Anal Biochem 66:372
298. Kato T (1978) Kagaku-To-Seibutsu 16:336
299. Schulman MP, Gupta NK, Omachi A, Hoffman G, Marshall WE (1974) Anal Biochem 60:302
300. Cox C, Camus P, Buret J, Duvivier J (1982) Anal Biochem 119:185
301. Yeung KK, Carrico RJ, Christner JE, Boguslaski RC (1978) Measurement of ATP and ligand-ATP conjugates by enzymic cycling with co-immobilized hexokinase and pyruvate kinase. In: Broun GB, Manecke G, Wingard LB (eds) Enzyme engineering, vol 4. Plenum Press, New York, p 427
302. Cremonesi P, Strada D, Carrea G, D'Angiuro L (1983) J Appl Biochem 5:59
303. Aizawa M, Wada M, Suzuki S (1980) J Solid-Phase Biochem 5:35
304. Huck H, Shelter-Graf A, Danzer J, Kirch P, Schmidt HL (1984) Analyst 109:147
305. Murachi T (1980) Bitamin 54:503
306. Sakaguchi Y, Sugahara M, Endo J, Murachi T (1981) J Appl Biochem 3:32
307. Pau CP, Rechnitz GA (1984) Anal Chim Acta 160:141
308. Wallace TC, Coughlin RW (1978) Biotech Bioeng 20:1407
309. Shubert F, Kirstein D, Schroder KL, Scheller FW (1985) Anal Chim Acta 169:391
310. Yao T, Musha S (1979) Anal Chim Acta 110:203
311. Carrico RJ, Christner JE, Boguslaski RC, Yeung KK (1976) Anal Biochem 72:271
312. Nicolas JC, Chaïnreuil J, Descomps B, Crastes de Paulet A (1980) Anal Biochem 103:170
313. Harper JR, Orengo A (1981) Anal Biochem 113:51
314. Payne DW, Shikita M, Talalay P (1982) J Biol Chem 257:633
315. Malinauskas AA, Kulys JJ (1979) Biotech Bioeng 21:513
316. Pfeiffer D, Scheller F, Janchen M, Bertermann K (1980) Biochimie 62:587
317. Mizutani F, Shimura Y, Tsuda K (1984) Chem Lett, p 199
318. Kura MR, Wichmann R, Oden U, Wandrey C (1980) Biochimie 62:523
319. Katayama N, Urabe I, Okada H (1983) Eur J Biochem 132:403
320. Cha S (1970) J Biol Chem 245:4814
321. Miyawaki O, Nakamura K, Yano T (1982) J Chem Eng Japan 15:142

Subject Index

hydroxy acid   100
3-hydroxybutyrate   110
2-hydroxyisocaproate   110
hydroxyketone   110
β-hydroxysteroid dehydrogenase   115

immobilization of NAD   34
immobilized blue-green algae   77
immobilized chloroplast   73
immobilized enzyme   32, 87
immobilized green algae   77
immobilized microbial cell   79
immobilized photosynthetic organella   73
immunoassay
    electrochemical   48
    voltammetric   48, 49
    with coenzyme label   125
immunoglobulin   27
in situ regeneration   136
indicator reaction   123
insulin   26
isotope-labeled compound   113

12-ketochenodeoxycholic acid   113

lactase   87
D-lactate   109
L-lactate, cycling assay   125
lactate dehydrogenase see LDH
3S lactone   115
Langmuir-Blodgett   71
LDH   93, 97, 100
L-leucine   113, 139
lipoamide dehydrogenase   89
liquid crystal membrane electrode   16

malic enzyme   93
mean residence time   147
mediated oxidation   96
mediator   29, 57
membrane reactor   139
methane monooxygenase   114
methyl viologen   89
4-methyl-L-glutamic acid   113
modified electrode   29
monomer bearing NAD   132
myoglobin   20

NAD   87
    casein-bound   132
    chemically modified   128
    Dextran-bound   129, 131, 132
    dimer   89
    dimerization   16
    immobilization   34
    kinase   120

Sepharose-bound   129
    polyethyleneglycol-bound   131, 139
    polyethyleneimine-bound   131
    polylysin-bound   131
NAD-bearing copolymer   130
NAD-HLADH conjugate   110
NADH
    electrochemical oxidation   16
    electrolytic regeneration   35
    oxidase   95, 97
    peroxidase   95, 97
NADP   87
    chemically modified   133
    Dextran-bound   133
    enzymatic process for   120
    macromolecule-bound   133
    polyethyleneglycol-bound   134
    polyethyleneimine-bound   133
    polymer-bound   16
NADP-linked ADH   92
$NAD(P)^+$
    electrochemical reduction   14
    enzymatic regeneration   94
    process with regeneration of   115, 116
NAD(P)H
    enzymatic regeneration   90
    process with chemical regeneration of
        102
    process with electrochemical regeneration
        of   102
    process with enzymatic regeneration of
        104
    process with regeneration of   100
NAD(P)H-FMN reductase   97
3-β-naphthoyl-Nile Blue   126
Nernst equation   3
nicotinamide adenine dinucleotide see NAD
nicotinamide adenine dinucleotide
    phosphate   see NADP
nitrogenase   13
2-norbonanol   112

optical immunosensor   66
optically transparent electrode   5
optrode   65
organic conductor electrode   17
overpotential   89, 96
oxidation-reduction potential   89, 93
oxidative phosphorylation   100
oxygen electrode   46

partially-contained system   137, 139
PAPS   88
penicillin amidase   87
penicillinase   43
peroxidase   21, 27
D-phenyllactate   109